T0135976

Björn Schwerdtfeger

Pick-by-Vision:
Bringing HMD-based Augmented Reality
into the Warehouse

Logos Verlag Berlin

λογος

Bibliografische Information der Deutschen Nationalbibliothek

Die Deutsche Nationalbibliothek verzeichnet diese Publikation in der Deutschen Nationalbibliografie; detaillierte bibliografische Daten sind im Internet über http://dnb.d-nb.de abrufbar.

ISBN 978-3-8325-2627-6

Logos Verlag Berlin GmbH
Comeniushof, Gubener Str. 47,
10243 Berlin
Tel.:   +49 (0)30 / 42 85 10 90
Fax: +49 (0)30 / 42 85 10 92
http://www.logos-verlag.de

To You

Abstract

The technology of Augmented Reality exists for several decades and it is attributed the potential to provide an ideal, efficient and intuitive way of presenting information. However it is not yet widely used. This is because the realization of such Augmented Reality systems requires to solve many principal problems from various areas. Much progress has been made in solving problems of core technologies, which enables us now to intensively explore the development of Augmented Reality applications.

As an exemplary industrial use case for this exploration, I selected the order picking process in logistics applications. This thesis reports on the development of an application to support this task, by iteratively improving Augmented Reality-based metaphors. In such order picking tasks, workers collect sets of items from assortments in warehouses according to work orders. This order picking process has been subject to optimization for a long time, as it occurs a million times a day in industrial life.

For this Augmented Reality application development, workers have been equipped with mobile hardware, consisting of a wearable computer (in a back-pack) and tracked head-mounted displays (HMDs). This thesis presents the iterative approach of exploring, evaluating and refining the Augmented Reality system, focusing on usability and utility. It starts in a simple laboratory setup and goes up to a realistic industrial setup in a factory hall. The Augmented Reality visualization shown in the HMD was the main subject of optimization in this thesis. Overall, the task was challenging, as workers have to be guided on different levels, from very coarse to very fine granularity and accuracy. The resulting setup consists of a combined and adaptive visualization to precisely and efficiently guide the user, even if the actual target of the augmentation is not always in the field of view of the HMD.

A side-effect of this iterative evaluation and refinement of visualizations in an industrial setup is the report on many lessons learned and an advice on the way Augmented Reality user interfaces should be improved and refined.

Zusammenfassung

Obwohl die Technologie der Erweiterten Realität (ER) seit mehreren Jahrzehnten existiert und ihr eine erhebliches Potential als ideale, effiziente und intuitive Methode zur Informationsbereitstellung nachgesagt wird, findet sie noch keine breite Anwendung im industriellen Umfeld. Dieses ist primär darin begründet, dass für die Umsetzung eins ER-Systems viele grundlegende Probleme aus diversen Disziplinen gelöst werden müssen. Da in letzter Zeit in den Kerntechnologien diverse Fortschritte erzielt wurden, ist es uns nun möglich die Entwicklung von Anwendungen auf Basis dieser Technologie zu explorieren.

Als exemplarischer Anwendungsfall wurde der Kommissionierprozess gewählt. In dieser Arbeit wird hierzu über den iterativen Prozess des Design und der Evaluation von Methoden der Erweiterten Realität zu Unterstützung der Kommissionieraufgabe berichtet. Unter der Kommissionieraufgabe wird das Zusammenstellen/-suchen von Teilmengen auf Basis eines Auftrages aus einer Gesamtmenge verstanden. Die Kommissionieraufgabe ist bereits Gegenstand vieler Optimierungen gewesen, da sie täglich millionenfach im Industriealltag durchgeführt wird.

Für die Entwicklung der ER-Kommissionieranwendungen wurden Arbeiter mit mobiler Hardware ausgestattet. Diese besteht im Wesentlichen aus einem kleinen Tablet-PC, der in einem kleinen Rucksack getragen wird und einer Datenbrille die mit einem Positionserfassunssystem gekoppelt wurde. Diese Arbeit präsentiert sodann den iterativen Ansatz des Explorierens, Evaluierens und Präzisierens de ER-Systems mit dem Fokus auf Benutzbarkeit aber auch auf Nützlichkeit. Die Entwicklungen und Evaluationen dieser Arbeit wurden Anfangs in einem kleinen Laboraufbau gestartet und später in einer Versuchshalle unter realen Bedingungen fortgesetzt. Hauptsächlicher Gegenstand der Optimierungen waren die ER-Visualisierungen, die in der Datenbrille angezeigt werden. Die Herausforderung der Aufgabe besteht darin dem Arbeiter eine Navigation auf verschiedenen Stufen (von grob bis fein und genau) zur Verfügung zu stellen. Das resultierende System besteht aus einer kombinierten und adaptiven Visualisierung die den Arbeiter präzise und effizient führt, auch wenn das eigentliche Ziel nicht soffort im Sichtfeld der Datenbrille liegt.

Als Nebeneffekt dieser iterativen Evaluationen in industriellen Umgebungen ist eine Katalog mit Erfahrungsberichten und Anweisungen für das effiziente Durchführen von Benutzerstudien zur Verbesserung von ER-Benutzerschnittstellen entstanden.

Danksagung

This work would not have been possible like it is, if the following people would have been absent. For that reason I would like to thank them (preferably in their mother tongue).

Meine größter und wichtigster Dank, gilt meiner Doktormutter Prof Gudrun Klinker, Ph.D. Danke, für die Betreuung, die Diskussionen, die Freiheit und das Vertrauen in ein Projekt, dass es in der form gar nicht mehr gab. Danke auch, dass Du immer noch die menschliche Seite des Ganzen bewahrt hast und Dir für das Wohlsein deiner Doktoranden auch dein Schlaf nicht zu teuer war. Es hat mich immer wieder verblüfft, wie Du in dem exakt richtigen Momenten den entscheiden und wichtigen Gedankengang hatest. Many thanks to Prof. Bruce H. Thomas, Ph.D., who did not hesitate one second to take over the task of being the second supervisor and who gave me a lot of valuable feedback.

Ein großer Dank gilt meinen Hiwis und Diplomanden, die mich wesentlich in der praktischen Umsetzung und den Evaluation in dieser Arbeit unterstütz haben: Troels Frimor, der das erste und umfangreichste UbiDisplay-Picking System gebaut hat. Erwin Yükselgil, danke für die Erhebung detaillierten Wissens über IQ und Persönlichkeitsprofile der Menschen an unserer Universität and darüber hinaus. Peter Pan, der die Ringe verdünnt hat, und dabei ganz übersehen hat, wie genial das eigentlich war. Michael Stather, der eigentliche Programmierer hinter dem Ganzen.

Besonders möchte ich mich auch bei den Logistikern vom fml-Lehrstuhl bedanken. Allen voran natürlich bei Rupert Reif, der diese Arbeit ganz entscheidend geprägt hat und mich immer wieder mit realen Logistikern konfrontiert hat. Aber auch Julia Boppert, Dennis Walch, Michael Schedlbauer, Johannes Wulz und Max Meister gilt eine Dank. Insgesamt muss ich sagen auch wenn unsere Zusammenarbeit Anfangs zögerlich startete, hat sie doch letztendlich exorbitante Früchte getragen. Vielen dank auch an Prof. Dr. Willibald A. Günthner, dafür, dass Sie das gesamte Projekt unterstützt haben.

Further more I would like to thank Margarita Anastassova, for all the discussions, the insights and for extremely putting the utility in my focus.

Zudem möchte ich mich bei Johannes Tümler und Fabian Doil für die Unterstützung in bei den Langzeittests und auch in sonstigen Kommissionierangelegenheiten bedanken.

Ein ganz großer Dank, gilt meinen Kolegen aus dem Fachgebiet Augmented Reality: Pete Keitler, Manuel Huber, Daniel Pustka, Michael Schlegel, Christian Wächter, Dr. Marcus Tönnis, Dr. Martin Bauer, Dr. Florian Echtler, Benjamin Becker, Dr. Verena Broy, Dr. Katharina Pentenrieder, Simon Nestler, Patrick Maier, Markus Duschel, Eva Artinger, Amal Benzina und Tayfur Coscun. Vielen Dank für die schöne Zeit, die fachliche Unterstützung, die Hilfe bei unzähligen Aufbauten und natürlich wenn es mal brannte, die Toleranz der Labor-einehmenden Experimente und Demos.

Ein weiter Dank gilt der ART GmbH, für die immer hilfsbereite Unterstützung und Kooperation in all den Jahren. Ohne die Leihgaben wären die, so wichtigen, Tests in der Versuchshalle nicht möglich gewesen.

Ein Dank gilt natürlich auch meinen anderen Studenten, die zwar nicht am Pick-by-Vision beteiligt waren, aber die mich tatkräftig in den anderen Projekten unterstützt haben: Frieder Pankratz und Wolf Rödiger, vielen Dank für Euren Einsatz und die genialen Lösungen. Andreas Demmel, Daniel Muhra, danke, dass ihr mir so viel Arbeit für den Laser abgenommen habt.

Auch den Partnern von Vodafone R&D, der LFK GmbH und Clemens Kujawski möchte ich für die Unterstützung und die inspirierende Zusammenarbeit in den anderen Projekten danken.

Weit zurückblicken möchte ich auch Frank Götz und Gitta Domik danken, die mich damals zu frühen Studienzeiten für das Thema Augmented Reality begeisterten.

Viel Dank gilt auch meine Freunden für die ausdrücklich nicht fachliche Unterstützung, die eben so wichtig für die Erstellung dieser Arbeit war. Leslie Klein möchte ich ganz besonderes für ihre Hilfe und Geduld danken. Letztendlich möchte ich mich bei meiner Familie für ihre Unterstützung während meines bisherigen Lebens danken.

Overview

1 Introduction 1

Even though the idea of Augmented Reality has existed for more than 40 years, there has not yet been a technological breakthrough. This is because many (un-) solved problems from very different disciplines have to be addressed. The challenge is nothing more than to extend our complex reality and to generate something useful.

2 Domain Analysis 9

What is the Order Picking Process? What are the state-of-the art technologies? Pick-by-Vision: Supporting Order Picking with Augmented Reality.

3 Improving AR User Interfaces 21

Specific issues when evaluating Augmented Reality user interfaces. A report on lessons learned from several iterations of user-based evaluation in industrial environments. A new usability engineering strategy is proposed.

4 System 47

Details on the Pick-by-Vision software. Reasons for using specific displays, input devices and tracking systems.

5 Evaluations 57

The iterative process of (re-)designing and evaluating the Augmented Reality visualizations and the Pick-by-Vision system itself

6 The Final Pick-by-Vision Visualization 119

A description of the final Pick-by-Vision (AR) visualization. It consists of a frame to highlight the box which is supported by an adaptable tunnel.

7 Conclusion 127

An industrial Augmented Reality application was iteratively developed and evaluated. This included the design of a new metaphor for guiding users to objects outside their field of view. The lessons learned during all this evaluations led to a proposal of new ways to improve usability of AR systems.

A Appendix 135

Configuration Files

Contents

1 Introduction 1

  1.1 Motivation 2
  1.2 Augmented Reality 3
      1.2.1 Industrial Augmented Reality Applications 4
  1.3 Logistics Applications 5
  1.4 The Problem of Executing Industrial Augmented Reality Evaluations ... 6
  1.5 Outline 6
  1.6 Differentiation and Cooperation 8

2 Domain Analysis 9

  2.1 Order Picking 9
      2.1.1 State-of-the-Art Technologies 10
      2.1.2 Naming Conventions used in Warehouses 11
      2.1.3 The Order Picking Sequence 11
      2.1.4 Logistics Figures 12
  2.2 Related Augmented Reality Solutions 13
      2.2.1 Related Experiments Supporting Order Picking with HMDs 13
  2.3 Pick-by-Vision 17
      2.3.1 Definition: Pick-by-Vision 17
      2.3.2 Supporting Order Picking with HMD-based Augmented Reality .. 17

3 Improving AR User Interfaces 21

  3.1 Introduction to User Interface Design 23
      3.1.1 User-Centered User Interface Design 23
      3.1.2 Utility 24
      3.1.3 Usability 24
      3.1.4 Design Guidelines 26
      3.1.5 Problems of User Interfaces Evaluation and Usability Engineering for AR 27
  3.2 User Studies & Structured Experiments 29
      3.2.1 Fundamentals of Empirical Experiments 29
      3.2.2 Types of Experiments 31
  3.3 Running User- and Expert-based Studies for AR 33
      3.3.1 How to Set Up 33
      3.3.2 How to run the Session 37

    3.3.3   How to Analyze and Proceed with the Results 39
3.4 Usability / Utility Engineering Processes for AR 41
    3.4.1   Usability/Utility Engineering Processes: State-of-the-Art 42
    3.4.2   A refined Usability/Utility Processes 43
3.5 Acceptance of New Technologies 45
3.6 Conclusion 45

4   System   **47**
4.1 Software 47
    4.1.1   Pick-by-Vision System 47
    4.1.2   Warehouse Layout 48
    4.1.3   Warehouse Management System 49
    4.1.4   HMD Calibration Wizard 49
4.2 Hardware 50
    4.2.1   Input Devices 52
    4.2.2   Output Device - HMD 52
4.3 Tracking System 54
4.4 Warehouses 55

5   Evaluations   **57**
5.1 Exploring the Use of AR to Support the Order Picking Process 59
    5.1.1   Objectives 60
    5.1.2   Experimental Setup 60
    5.1.3   Results 63
5.2 Finding a Convenient Visualization and Likely Successful Users Of AR
     Systems 65
    5.2.1   Objectives 66
    5.2.2   Experimental Setup 66
    5.2.3   Results 71
    5.2.4   Conclusions 79
5.3 First Public Evaluation 81
5.4 Improving the Meta Visualizations 82
    5.4.1   Objectives 82
    5.4.2   Experimental Setup 82
    5.4.3   Results 86
    5.4.4   Conclusions 89
5.5 Bringing AR into a Real Warehouse - A Pretest 91
    5.5.1   Informal Evaluation and Adaption to the New Warehouse 91
5.6 First Comparative User Study in a Real Warehouse 96
    5.6.1   Objectives 96
    5.6.2   Experimental Setup 97

      5.6.3   Results 98
      5.6.4   Conclusions 101
  5.7 A Stress Test and Concluding the Final Visualization 103
      5.7.1   Objectives 104
      5.7.2   Experimental Setup and Hypotheses 104
      5.7.3   Results 106
      5.7.4   Conclusions 112
  5.8 Verification of the Final Visualization 114
      5.8.1   Results 114
  5.9 Final Presentation thus far 116
  5.10 Conclusions 116

6   The Final Pick-by-Vision Visualization   119
  6.1 The Final Visualization 119
      6.1.1   Aisle Finding – Coarse Navigation 119
      6.1.2   Box Finding – Fine Navigation 120
      6.1.3   Textual Information 123
      6.1.4   Accidental State Change 124
  6.2 Conclusions 124

7   Conclusion   127
  7.1 Contribution & Future Work 128
      7.1.1   Pick-by-Vision 128
      7.1.2   HMD and User Strain 129
      7.1.3   Usability & Utility Engineering 130
  7.2 Final Words 131

A   Appendix   135
  A.1 Visualization 135
  A.2 Shelf Setup 137
  A.3 Warehouse 138
  A.4 Orders 138

Introduction

Even though the idea of Augmented Reality has existed for more than 40 years, there has not yet been a technological breakthrough. This is because many (un-) solved problems from very different disciplines have to be addressed. The challenge is nothing more than to extend our complex reality and to generate something useful.

Die Idee ist gut, doch die Welt noch nicht bereit.
(engl.: The idea is good, but the world is not ready yet.)

TOCOTRONIC, GERMAN ROCK BAND

This work exists in the current form because we tried to support a simple task with a technology which allows something mostly known only from science fiction movies. It allows us to augment our real environment with virtual objects to provide information of any kind. It is therefore called Augmented Reality. The task to be supported was to show workers out of which box on a shelf they have to pick items. We thought that using this technology to support this task is like breaking a butterfly on a wheel. Nevertheless, we set up a system which displayed (via the use of special goggles) a virtual arrow pointing to the real shelf from which the worker had to pick an item. We consulted experts for the design and shape of the three-dimensional arrow. However, when assessing the system in a first user test, it simply failed dramatically. The users quite often did not pick from the box to which they were directed by the virtual arrow.

To this end we took up the challenge of improving this Augmented Reality system in such a way that it is able to guide workers safely, precisely and quickly in an industrial warehouse environment. This process which consisted of eight iterations of designing and evaluating (the visualizations and the system) is the main part of this work. During these evaluations, we realized that the current methods of user interface evaluations are only partially suited for evaluating Augmented Reality user interfaces. To this end we discuss the aspects to be considered when evaluating Augmented Reality user interfaces and propose a new usability/utility engineering process, as the second main contribution of this thesis.

1.1. Motivation

Even though it has existed for several decades and is said to have the potential to provide a competitive advantage for industrial applications in globalizing markets, Augmented Reality (AR) is not yet widely used. AR is seen as the technology which allows an intuitive, efficient and thus ideal way for the presentation and interaction with information in our natural environment. For that reason, AR might be applied to different industrial scenarios for example to speed up processes, prevent workers from making errors, or reduce training times [39, 51]. If Augmented Reality has so many benefits, why is it still waiting for its technological breakthrough? This is mainly due to the fact that for the realization of such Augmented Reality systems, many principal problems from various disciplines have to be solved. Much progress has been made in solving problems of core technologies, but a lot of problems remain in the application domain and the actual user interface. Most industrial Augmented Reality setups are only demonstrations of the technology and are content with augmenting some virtual arrows to highlight important places. This is partially due to the fact that perfect display hardware does not yet exist, but mostly because it has already taken so much effort to solve the basic technological problems and calibrating all the components, that no resources have remained for developing the actual application. As a result, Augmented Reality is still a technology looking for applications.

To this end, this work focuses on the development of an industrial Augmented Reality application and thereby gives priority to make the application usable and useful. This is achieved by following a user-centered development. We therefore integrated users and domain experts into the development process from the beginning of the project. We started with task analyses and field observations and we made intensive use of user-based evaluations of the system.

Our industrial use case is the order picking process of logistics applications. In such order picking tasks, workers collect sets of items from assortments in warehouses according to work orders. This order picking process has been subject to optimization for a long time, as it occurs a million times a day in daily industrial life and, minimal improvements can thus provide a huge benefit. To this end, we investigate the application of Head-mounted Display (HMD)-based visualizations (both AR and non-AR) to support the order picking process. This HMD-based support of the order picking process has been named "Pick-by-Vision" [52].

A byproduct of this iterative evaluation and refinement of visualizations in an industrial setup is the report on many lessons learned and some advice on the way Augmented Reality user interfaces should be improved and refined.

In the remainder of this chapter we will introduce the technology of Augmented Reality, explain the requirements of modern logistics applications, introduce the problem of evaluating industrial AR applications and give an outline of this document.

1.2. Augmented Reality

The term Augmented Reality (AR) has had a number of definitions. However, the Augmented Reality community widely agrees with the definition by Azuma [6], which consists of the three summarized criteria: Augmented Reality 1) combines real and virtual, 2) is interactive in real-time, and 3) real and virtual world are registered/alligned in 3D. Two aspects of this definition are important to notice. Firstly, it does not only mean bringing computer graphics to the real world, but also includes, for example, augmented sound or augmented haptic feedback. Secondly, the aim of designing a good Augmented Reality interface is not necessary to make the virtual objects distinctive from the real objects, as the user often shall see the Augmentation (the objects we augment) immediately. This is for example the case in modern television sportcasts, where offside positions are virtually marked on the sports field, which is strictly not Augmented Reality because is not interactive in real-time.

Various ways for the realization of Augmented Reality systems exist, but the core components beside the actual application are a 3D world model, a tracking system, and a display device. These components are explained below.

3D World Model   A model is required which describes the spatial relationships of virtual and real objects. The model particularly describes where, in relation to real objects, the virtual objects shall be placed. Often further information about the real world is required, for example, to provide a correct depth perception by giving the illusion of real objects occluding virtual objects (see e.g. Fig. 5.4a). 3D world models are also important to prevent a navigation system from guiding users directly into a physical obstacle, instead of around it.

Tracking Systems   To be able to augment the real world with spatially aligned virtual objects, the virtual 3D world model needs to be aligned with the real world. That means the pose (position and orientation) of the system/user has to be estimated in relation to the real world. As the users' pose and/or the environment are typically not static the pose has to be updated continuously. In the virtual reality community the term *Tracking System* has been established for all systems that return the pose of an object with a certain frequency [9]. Common tracking systems use calibrated cameras and image processing algorithms to detect for example fiduciary markers or even features of real objects. Alternative systems make use of concepts such as time-of-flight measurements (like GPS), inertial sensing (gyroscope, accelerometer), or mechanical tracking. All tracking systems have their benefits and disadvantages. They can be characterized particularly by update-rate, latency, accuracy, operating range, robustness, line of sight required (between sensors), degrees of freedom and the amount and weight of equipment which is attached to the user/environment. A perfect atomic tracking system for Augmented Reality, which continuously (that means under all circumstances) delivers an accurate pose does not exist. That is why current systems tend to make use of hybrid setups by fusing the signals from different sensors [69]. This sensor fusion however, is not trivial as all sensors need to be calibrated to each other in spatial and time domains.

Displays   According to Azumas' criteria [6], the augmentation can be of any form, but most Augmented Reality systems only make use of visual augmentations. To generate such an augmented view, displays are required which provide a method to combine the view of the real world with virtual objects in the form of computer generated graphics. If we classify these displays by the technology and position where the augmented image is displayed, the following three classes exist: the combination is either done in the user's eye (retina projection / semi-transparent monitor – a.k.a. optical see-through display), in front of the users eye (overlaying a video with the computer graphics on a monitor – a.k.a video see-through display) or in the real world (projection on the real world/or on a special surfaces). Additionally, Bimber et al. [13] classify visual Augmented Reality displays by their spatial location in Head-attached (Head-mounted), Hand-held, and Spatial Displays. Initially, we have explored the use of all three kinds of Bimber's display types in our usage scenario [123, 40], but this thesis only focuses on head-mounted displays.

Augmented Reality has not yet had its public breakthrough, but probably will have it soon according to the Gardner Hype Cycle 2008 [47]) and several articles in newspapers from 2009. The first end user applications have already appeared by the end of 2009 in Apple's App-Store for the iPhone or Googles' Android Market Place for the G-Phone [5, 1, 88, 91, 49]. The iPhone and other new generations of smart-phones are currently hyped as ideal devices for the realization of many kinds of Augmented Reality applications, as they combine advanced tracking technology and a display in a mobile device. Most of these applications do not consistently comply with Azumas' criteria [6], e.g. they are just context aware applications, which overlay 2D labels on videos. However, these "pseudo Augmented Reality" applications are probably the ones which will publicize the concept of Augmented Reality. The general public will soon define as Augmented Reality everything which mixes views of reality and virtuality in 2D or 3D or without proper alignment. At this point, the well known mixed reality continuum description by Milgram et. al. [89] should be mentioned. One end of this continuum is the real world, and the other is the complete virtual world. The transition from one end of the continuum to the other goes via Augmented Reality and Augmented Virtuality.

1.2.1. Industrial Augmented Reality Applications

Augmented Reality applications have been proposed for various kinds of areas, including medical, entertainment, marketing, military, navigation, sight seeing and industry. Due to the scope of this thesis, we will give a short overview of the industrial applications. A lot of research has been done in this area. Early work with HMD-based AR was performed by Mizell et. al. [90]. They developed a system to support the assembly of wire bundles for airplanes. In a field test to assess the performance of the system, they had rather pragmatic problems: Users more often chatted with passers-by about the "fancy" HMD rather than working on the task. In addition to that, some users did not accept the system because the bulky HMDs were hair-mussing. The German government founded three large Augmented Reality related cooperations between research institutes and industry. The first project was called ARVIKA [39] and its successors are ARTESAS and AVILUS. The main fields of application are service and maintenance, production and assembly, and factory planning. Even though a lot of work has been done in this area, industrial applications rarely exist. Regenbrecht et. al. [113] analyzed 10 industrial Augmented Reality projects, which were executed within 5 years and conclude that the development of real industrial Augmented Reality applications is challenging. There are only a few systems which are in real industrial use, whereas non of them uses an HMD. One of these productive systems is for example an intelligent welding gun [34] used in experimental vehicle construction. A small display mounted on a welding gun is used to guide the worker to the welding spot. Another productive system is used in Augmented Reality based factory planning [108]. Photographs of the factory are analysed and augmented on stationary computers.

Further applications, which are directly related to our work will be presented in Sec. 2.2.

1.3. Logistics Applications

The basic conditions in the field of logistics have changed rapidly over the last years, with the market demanding customized products. As an example, 20 years ago automotive manufacturers offered three model series, while nowadays they are offering nearly ten. This is accompanied by an increase in the variants within one series. Thus, production and logistics systems have to become *supra-adaptive*[1] [51]. This requires that production processes as well as supporting IT systems are designed in a manner that enables workers to quickly handle new working conditions and environments. Thus the need exists to make workers improve under such working conditions, without increasing their stress level, and while preventing them from making errors. This requires systems to support the worker with just the right information at exactly the right time. Such supporting systems have to provide detailed working instructions which have to be presented in a highly intuitive and precise way. As a result, workers can then start executing arbitrary jobs efficiently and without errors – and with minimal prior training. Augmented Reality is judged to have the potential to provide this very functionally.

---

[1]to be able to adapt with minimal effort to global dynamic changes

1.4. The Problem of Executing Industrial Augmented Reality Evaluations

Several obstacles exist when trying to apply AR to real life scenarios. In fact, most AR systems thus far have remained laboratory prototypes, mainly due to the lack of adequate hardware, but also due to not yet resolved usability issues [86]. Furthermore, we have to determine the situation in which the user can benefit from AR, without knowing how good our AR user interface is. Thus, we must verify usability to determine whether the system is effective [86]. This is a difficult proposition as we develop user interfaces at the limit of what is known or common practice in our field.

This goes along with the problem that the utility of an AR application heavily depends on how well it fulfills the user's needs. But in order to be able to elicit such user needs from future users, we need high-fidelity AR prototypes [4]. Mature AR systems help elicit many more concrete user needs than low-fidelity prototypes.

Due to these facts it is challenging to develop and evaluate such new AR user interfaces and applications. The information visualization community has similar problems and thus promotes the reporting of long-term usage and field studies in natural settings and the investigation of new evaluation methods [110]. Furthermore, the execution of experiments in an industrial environment under realistic conditions with real tasks is complex, time consuming, and expensive – especially when incorporating real factory workers. It is still an open question how to execute such user studies in the field of AR. Dünser et al [32] state that, until now, it is partially unclear how to analyze the different user-related questions. Additionally, the funding of such experiments usually comes from industry partners who expect results in a meaningful and summative manner in a short amount of time. But the development of today's AR systems has not yet sufficiently matured, that well-developed and perfect systems are rarely available. We therefore have to compare our immature AR prototypes with established technologies in summative experiments. Naturally most AR systems fail this objective test, because the prototypes are affected by usability problems and do not fullfil all the relevant user needs. Finally we have to be careful when executing and interpreting such experiments, as we have to give incentives for further investment.

1.5. Outline

This work contributes a Head-mounted Display-based Augmented Reality system to guide workers in a warehouse – a Pick-by-Vision system. The Augmented Reality visualization, guiding the user to the right box on the right shelf, was the main subject of optimization. This thesis contributes detailed insights into the long and iterative process of developing, evaluating and refining the Pick-by-Vision system. This process started in a "simple" laboratory setup and proceeded to a realistic industrial setup in a factory hall. While performing these evaluations, we realized that we had to adapt the traditional methods of executing experiments to the requirements of evaluating AR user interfaces and we thus present a discrete chapter on improving AR user interfaces in general. This work consists of the following chapters.

Chapter 2 – Domain Analysis In this chapter the order picking process and state of the art technologies to support it are analyzed. This includes a task decomposition, related work and a discussion on how this task can be supported with Augmented Reality.

Chapter 3 – Improving AR User Interfaces This chapter discusses what needs to be considered when aiming to improve usability and utility of Augmented Reality user interfaces via user- and expert-based evaluations. This discussion basically reflects the experiences we had during all the studies of this work on the traditionally used methods. In particular, it takes the lack of expert knowledge in the unexplored design space of Augmented Reality user interfaces into account and thus proposes a new usability/utility engineering strategy.

Chapter 4 – System The technical aspects (Hard-/Software) of the *Pick-by-Vision* system are presented. However, a discussion on suitable wearable computing hardware and tracking systems is not within the scope of this work.

Chapter 5 – Evaluation This chapter presents the eight consecutive user studies we executed to improve the Pick-by-Vision system. Overall we have gathered data from more than 100 subjects picking altogether more than 10,000 items out of different boxes in different warehouses, guided by one of our various AR-based visualizations. The series of experiments started in a lab with small experiments to solve basic usability problems and ends up in a realistic warehouse with two-hour stress tests, to optimize the visualisation and gain insights into the longterm effects of using the system.

Chapter 6 – Final Pick-by-Vision Visualization The Pick-by-Vision visualization which resulted from the iterative evaluations is presented in this chapter. The visualization is displayed in the tracked HMD and makes use of a combined Tunnel and Frame paradigm to guide workers in a warehouse.

Chapter 7 – Conclusion The last chapter summarizes the results and makes final conclusions.

1.6. Differentiation and Cooperation

This work was partly done in cooperation with two other Ph.D. Students. Reif [114] evaluated the general application of AR and non-AR head-mounted display based systems to support the order picking process. In his work, he basically discussed the efficiency of such systems and the different logistical aspects. The experiments of Sec. 5.6 and Sec. 5.7 were done in cooperation with this work. His work focuses on the argumentation for stakeholders' investments. Tümler [145] focused among other things on the analysis of user strain when working with HMDs over longer time periods. The strain analysis in the experiment described in Sec. 5.7 was inspired by him and executed with his cooperation [127].

Domain Analysis

What is the Order Picking Process? What are the state-of-the art technologies? Pick-by-Vision: Supporting Order Picking with Augmented Reality.

The development of the Pick-by-Vision system started with an analysis of the order picking task. We did this to understand the environment in which our actions take place and to elicit the important user needs. Initially we spoke to experts, did literature reviews and went to several logistics companies, looked at different workplaces and interviewed workers and foremen. After the first prototype was finished, we brought the domain experts and workers to our lab to observe them working with our system (as described in Chap. 5). This analysis took place in close cooperation with the Institute for Materials Handling, Material Flow, Logistics from TUM [114].

This chapter firstly presents an analysis of the order picking process and state of the art technologies to support it. After that a review of related Augmented Reality solutions to support such tasks is given. Finally we describe the important subtasks of the order picking process and how they can be supported with Augmented Reality.

2.1. Order Picking

Order picking is the gathering of goods out of a prepared range of items following customer orders [149]. A simple order picking system can be compared to shopping in a supermarket. One picks the goods from the shelves in accordance with a shopping list. In the automotive industry, order picking is used to collect parts that are required for a customer's specific car. After picking these items into a basket, the basket is brought to the corresponding car, where the parts are mounted to the car.

Flexibility and fine motor skills of human beings are needed to actually grab the different items. Furthermore the product range, and thus the variety of items, steadily increases while, in contrast, the size of orders is decreasing. Thus, complete process automation is not the appropriate answer and humans are often the best solution for order picking [50]. The amount of work in a warehouse can highly vary with the season or due to other reasons [150]. This results in the fact that order picking is often performed by temporary workers. Those workers have to be supported in such a way that they can easily start an efficient and error free job, while minimizing the training period.

When speaking about order picking in this work, we basically address man-to-goods setups. These are setups, in which workers physically move through warehouses with different shelves in different aisles and pick goods out of boxes. Even if principles and solutions are transferable to other setups like goods-to-man they are left out here. A discussion about different warehouse setups and the applicability of Augmented Reality technologies can be found in [114].

2.1.1. State-of-the-Art Technologies

Different techniques exist to support the order picking in warehouses [50]: Conventionally, workers execute their orders with paper lists which are intuitive for human beings but laborious to handle. In this work, we call those systems *Pick-by-Paper* systems. Such a list can be seen in Fig. 2.1.

Picking Order

Order Number: 21
Page: 1 of 1

Order Line	Storage Location	Article Number	Amount	Acknowledgement
1	2 E 03	85668707	1	O
2	2 B 03	52158728	2	O
3	3 C 05	123645	1	O

Date & Signature:

Figure 2.1.: A typical picking order list, as used in Pick-by-Paper systems.

Modern order picking systems are *Mobile Data Terminals*, *Pick-by-Voice* or *Pick-by-Light* systems [50] (compare Fig. 2.2). All these technologies have specific advantages as well as disadvantages. Pick-by-Voice supports the worker by providing all instructions through the computer's speech output and is controlled via voice recognition. Unfortunately, Pick-by-Voice systems face difficulties in noisy industrial environments. Furthermore, it is questionable whether the order picking worker likes being bossed by a monotone voice the whole day. Pick-by-Light offers visual aid for the worker by installing small lamps in combination with simple digital displays (stating the amount of items to pick) on each storage compartment. Pick-by-Light systems have the problem that the displays have to be elaborately integrated into the shelf construction and thus are very expensive and inflexible. Furthermore such setups only allow one worker per aisle. Pick-by-Light is most suitable for order picking stations with a high throughput. Brynzer et. al. [22] state that better user interfaces could speed up the order picking process.

As well as the widely established systems, several customized solutions and research applications exist: One concept is *Pick-by-Point* [87] wich uses a moving light source to mark the storage location. Another concept which is similar to that and currently under development by Technische Univeristät München is *Pick-by-Laser*. It makes use of an Augmented Reality laser projector [126], which is mounted on a picking trolley.

Figure 2.2.: Common order picking technologies: MDT with scanner (1), Pick-by-Light (2) and Pick-by-Voice (3). Courtesy of [127].

2.1.2. Naming Conventions used in Warehouses

A simple but effective method for the naming of item locations is to alternate numbers and characters, e.g. : a number for the shelf, a character for the layer and a number for the column [114]. This results, for example, in names like "3 A 13" (3rd shelf, first layer, 13th column), which less often produces twisted numbers than using for example: "3 1 13". Warehouses are often filled up in a – *chaotic* – fashion instead of related articles being placed close to each other. This makes it possible to use every storage location. The warehouses in the experiments (see Chap. 5) were always filled up in this way. This prevented test participants from remembering the structure of the assortment and thus confounding the experiment.

2.1.3. The Order Picking Sequence

Before describing the actual order picking sequence, some terms have to be introduced. A *picking order* consists of a list of *order lines* (or *positions*). Each order line contains one type of *item* (or *article*), and the *amount* of items to be picked of this type. In our setups the order line also consisted of an *article number*. The article number is given for quality assurance reasons. It has to be compared with the number printed on the picked article.

Such a picking order is executed in the following sequence. At the beginning of each order, the worker needs to get the list of items to pick (the order list). In the case of using a Pick-by-Paper system, the worker moves physically to the station where he picks up the order list from a stack. In the alternative case of using one of the electronic picking systems, he simply requests the next order via his wirelessly connected device (e.g. the portable computer running Pick-by-Vision). Thereafter the worker has to pick up an empty collecting unit (and, in most cases, he puts it on a picking trolley). During the actual picking process, the worker handles the order lines from his list sequentially. For that, the worker needs to navigate to the right box in the right aisle/shelf. This task consists of a coarse navigation (finding the right aisle/shelf) followed by a fine navigation (finding the right box in this shelf). Having navigated to the right box, the worker needs to check the number of items to pick. In most logistic processes the worker has to make sure that he picks the correct item, by following a control mechanism. In our scenarios, we mostly used a comparison of the article number on the order line with that on the label of the item. To finish an order, the worker delivers the collecting unit to a station. In the case of Pick-by-Paper, the list has to be signed and must be put into the collection unit.

2.1.4. Logistics Figures

The quality of an order picking system is mainly measured according to the time and error rate. These logistics criteria will now be presented.

Order Picking Time

The order picking time is divided into four interleaved tasks[114]. The *base time* includes all tasks at the beginning and the end of one order (e.g. login at the system, pick-up and delivery of the collecting unit or the paper list). The *way time* is the time the user needs to physically move through the storage area, and the *picking time* consists of actually grabbing the item with the following delivery in the collecting unit or on a conveyor. Finally, the *dead time* includes the search for information, human information processing and all process steps that are not necessary for the actual task (e.g. opening the packages). The measure to compare the performance of order picking setups and systems is order lines per hour, rather than the overall number of items per hour. Order picking systems mainly focus on the reduction of the dead time, but also on the base time, because of an online link to the Warehouse Management System.

Order Picking Errors

Errors have a strong influence on the quality of delivery and the relationship between clients and suppliers. Thus, zero-defect picking is one major goal in optimizing order picking processes [50]. Picking errors can be divided according to their causes: a) wrong amount, b) wrong item/article, c) order line/missing article (wrong amount), d) damaged article. For Pick-by-Vision as an Augmented Reality system, the most interesting error is a wrong item because this is caused by picking out of the wrong stock location. This means that the visualization does not clearly highlight the right stock location. With logistics tasks the system must also avoid picking the wrong amount and missing articles.

The number of errors depends on the entire number of picked items. This is called the error rate. Among other things, the error rate depends on the order picking technology. The error ratio is typically the quotient of wrong order lines per overall picked order lines given in percentage. For a Pick-by-Paper it is normally 0.35%, for Pick-by-Light 0.40% and Pick-by-Voice 0.08% [140]. However, each error needs to be prevented, because it often leads to high follow up costs. For example, in the automotive industry each wrongly picked item can lead to halting the production line. The application of Augmented Reality for the order picking process primarily focuses on active error prevention, by highlighting the storage location to pick from in the best possible way. Reactive error prevention techniques [51] detect wrongly picked items, for example by making use of scanners, laser curtains in front of the shelves or weight checkers.

2.2. Related Augmented Reality Solutions

Several solutions which make use of Augmented Reality were proposed to support order picking like search tasks. One concept is to use *I/O Bulb*-like [148] setups. These are atomic units that perform (near-) coincident projection and video acquisition. Such setups are by now known as *projector-camera devices*. Butz et al. [24] use such a device to project onto points of interest in the environment. They estimate the position of the items by optical square marker tracking. Raskar et al. [112] have a similar device, but make use of a variant of RFID tags that are attached to the items in warehouses and transmit geometric data of the items.

Reitmayr et Schmalstieg [116] explored HMD-based navigations by using augmented sign posts, world in minature and compass like metaphors. We discuss further approaches to navigate users with tracked HMDs in Sec. 2.3.2.

2.2.1. Related Experiments Supporting Order Picking with HMDs

Some research has been performed before and during the time of this thesis to study the use of HMDs in order picking like scenarios. Most of the work investigates the exploration of economic benefits of using HMDs for mobile information presentation. Two of the systems presented below make use of HMDs displaying low-fidelity Augmented Reality visualizations to support the order picking process. One of these projects did not focus on the visualization but rather on the strain produced by the longterm usage of Augmented Reality. The following research projects were investigated.

Mobile Display: Cockpit Assembly Alt [2] conducted a user study to explore the economical potential of HMD-based systems. On that account users were equipped with an HMD showing context sensitive but not AR visualizations. The system control was realized via a voice input. This system was evaluated in a car cockpit assembly scenario. The cockpit modules were mounted on a workbench and the parts and tools were mostly located on a shelf behind the user. Users were guided to the item locations by a kind of non-AR visualization (which is not described in detail). The visualization was accompanied by textual descriptions and images of the article. Ten users had to execute a task consisting of six steps three times. In comparing the task completion time when using the HMD with the time users need when utilizing paper based information, the former group of users performed about 26% faster. Alt [2] states that most benefit comes from the concurrent gathering of information while performing other actions (e.g. walking).

Mobile Display: Field Study Truck Assembly – Followed by a Lab Test Brau et al[19] ran a four week long field test concerning the assembly of trucks. Within this process the assembly workstation has to be supplied with items according to the Just-in-Time and Just-in-Place principle. Orders are highly diverse and contain between 80 and 120 positions. Currently, this process is supported by long paper lists attached to the picking trolley. For the field test, 13 voluntary workers were equipped with a mobile display (the Nomad HMD – Sec. 4.2.2) showing a scrollable order list. Users had to work a full 40 hour work week including the typical 45 minute lunch break. The largest challenge was to ensure the permanent use of the system including the process integration and an error free user interface. The users complained about discomfort caused by wearing the system. Some also suffered from headaches or had sore eyes. Due to that, several participants terminated the experiment before the end of the test. For that reason a roll-out of the system was not recommended.

As a result of this field study, a lab test was executed to gain more detailed insights into the longterm usage of the Nomad HMD [76, 41]. To this end 45 users were randomly separated into three groups (inbetween-subject design). The first group got instructions via printouts; the second group received the same printouts, but had to wear an HMD which was switched off; and the third group got exactly the same information via the HMD. The users had to work for one day (7.5 hours plus two breaks of 30 and 15 minutes) on three different tasks: a) order picking, a) pre-assembly of fuse boxes and c) assembly of construction toys.

The test was accompanied by intensive observation methods from the field of occupational medicine and psychology. Several examinations of the visual system were performed. Before the test, examinations were executed to ensure that only users with perfect eye sight would participate. After the test an examination was executed to look for changes in the visual system – no changes were found. A comparison of subjectively measured strain (questionnaire) or objectively measured strain (heart rate, heart rate variability) showed no difference between the three groups. Some users complained about headaches when wearing a disabled HMD, which was related back to psychological reasons. About 20% of the users mentioned problems with the visual system after having used the enabled HMD. They complained about a pressure in their eyes. This particularly appeared during the assembly of construction toys, where the users had to concentrate on the instructions displayed in the HMD more frequently than during the other tests. Users who were instructed by the HMD made more mistakes and were slower with both assembly tasks. The researchers related this back to the poor readability of the information shown in the display – the visualization was optimized according to the criteria by Oehme [104]. In contrast, the users performed better in the order picking task when being supported by the HMD.

Mobile Display: Field Test Order Picking   Reif et al. [115] performed another short term test in a real warehouse, comparing a Pick-by-Paper system with a Nomad HMD (compare Sec. 4.2.2) used as mobile display to support the order picking process. The HMD was used to display the order list and the system control was done via voice-entry. The voice-entry could be operated with 20 words and did not require training for the individual users. This fact was stated to be very important for an industrial application with changing workers (especially temporary workers) or for an evaluation with several test persons. Every command was confirmed with the same simple speech input except for the confirmation of the picked items, which had to be acknowledged by repeating the amount. Because of the small number of speech commands, the user became familiar with the system within a short time span. The test was performed in a temporary empty warehouse, which consisted of eight shelves with four aisles and a total of more than 600 stock locations. Each of the 14 test participants had to pick 125 items with each technology (within-subject design). The user performed only slightly faster with the HMD-based system. Yet a large learning effect was found. The error rate was much better for the HMD-based system (0.12%) in contrast to the Pick-by-Paper system with (0.84%).

After this test, we ran two experiments in cooperation with the authors of this system. The experiments are described in Sec. 5.6 and Sec. 5.7. Reif [114] mainly focused on the logistics aspects, the return on investment analysis, and utility, whereas we were primarily interested in the general use of the system under real conditions and, in particular, in the usability of the visualizations.

Strain Test: Augmented Reality vs. Pick-by-Paper Tümler et. al. [146] compared an Augmented Reality based order picking system with a Pick-by-Paper system. This evaluation focused on user-related issues in long-term use, mainly measuring strain by analyzing "heart rate variability" (HRV) during a two hour work phase. They tested 12 subjects in a within-subject design in two sessions. The tests for an individual subject were performed on another day at the same time of the day, as the heart rate variability is confounded by the time of the day. The warehouse consisted of a setup of six shelves (no real aisles) with a total of 58 boxes. The Augmented Reality order picking system was at a rather immature development stage. The system furthermore had a "huge" latency of 0.3 seconds, because it was implemented on a UMPC. The system made use of the Nomad HMD (compare Sec. 4.2.2) and optical tracking markers located under each box. The system control was realized via the spacebar of a mini keyboard mounted on the forearm. The visualization worked in the following way. When the actual box to pick from was not within the field of view of the HMD, a 2D arrow pointed in the direction of the box. When the box to pick from was within the field of view, a square highlighted the tracking marker which was mounted under each box. They concluded that two hours of continuous work leads to stress without and with AR. However, they did not observe an increased strain when using an Augmented Reality system compared to the Pick-by-Paper system. The users performed about 30% slower using the AR system, which the authors relate back to the low fidelity of the prototype. Nevertheless the error rate was less for the AR system, being 0.3% compared to 0.92% for the Pick-by-Paper System. Some users complained about headaches, which were mostly related back to a wrongly adjusted headband. Other users complained about sore eyes.

In the experiment we describe in Sec. 5.7, we try to reproduce the results of this experiment, in cooperation with the authors.

Explorative Test: Augmented Reality vs. Pick-by-Paper Mück et al. [93, 28] also compared a low fidelity Augmented Reality based order picking system with a Pick-by-Paper system. They explored the use of the system in an artificial warehouse with relatively large storage locations (stacked tables) consisting of two aisles with five columns and two rows. The system was based on a video-see through HMD and made use of optical square marker tracking. The user interface displayed several redundant pieces of information for guiding the user to the strorage location: a 2D floor plan in combination with a level meter, highlighting of the next item location, a textual description of the storage location. For showing the actual storage location with an augmentation, they used a transparent filled rectangle in front of the box. They observed a huge learning effect for the AR system and a better performance for AR (compared to Pick-by-Paper), but also state that most participants had orientation problems. The users rated the floor plan as best navigation aid (60%), followed by the textual description (35%) and the AR visualization (20%). Further details on the experiment (e.g. the number of participants) are missing in the description of the experiment.

2.3. Pick-by-Vision

We will now present what to consider when designing Augmented Reality visualizations to support the order picking process.

2.3.1. Definition: Pick-by-Vision

In the analogy to the state-of-the-art systems, all systems using HMDs to support the order picking process were labelled *Pick-by-Vision* systems by Günthner et al.[52], as they primarily provide information via the visual sense. However, we have to further subdivide this definition, as we can use an HMD as a simple mobile information display or as a real Augmented Reality display. The first setup, which we name *Pick-by-Vision (2D)* in particular does not require the use of tracking technologies to estimate the user's position, and thus presents the picking information as textual information in the form of a list of items or images, etc. We call other systems, which use tracking and make explicit use of AR visualizations, *Pick-by-Vision (AR)* systems.

2.3.2. Supporting Order Picking with HMD-based Augmented Reality

We described the different steps from requesting and order until the delivery of the picked items in Sec. 2.1.3. However the key steps of this process, to be supported with augmented reality, are the navigation to the right box in the right aisle/shelf. At first we have to guide users to the right shelf (the *coarse navigation*), followed by a second visualization (the *fine navigation*) to highlight the actual box to pick from. There is a third type of navigation which we also have to support: As we are dealing with HMDs and the problem of a small field of view, we must support the actual visualizations by a meta visualization to help users find the locus of augmentation. All three visualizations will now be discussed.

Meta Navigation

To direct the user's attention to objects that are currently outside the field of view of the HMD, we make use of a *Meta Visualisation*. The problem is illustrated in Fig. 2.3. The Meta Visualization, has two main attributes: 1) it guides the user to objects of interest outside the field of view and 2) it is always in the field of view (when needed). The development of a suitable Meta Navigation to guide users in 4π steradian was one

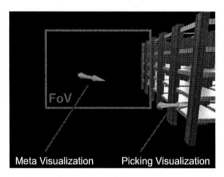

Figure 2.3.: Concept of a Meta Visualization: What the user needs to see is the actual picking visualization (the orange arrow). As the field of view of the HMD is limited (e.g. the red square), we need a meta visualization (as an example the green arrow), which guides the user to turn toward the object since it is currently outside the field of view of the HMD.

of the main subjects to be optimized within this work (see Chap. 5). This problematic issue of guiding a user to objects in arbitrary places when wearing a tracked HMD has been rarely adressed thus far. Early work was done by Feiner et. al. [37]. They guided the user's attention to objects outside the current field of view with a *rubber band-*like dotted line visualization, combined with highlighting the object. Höllerer et al [65] display a cone-shaped 2D pointer in a tracked HMD to point to dedicated out-of-sight landmarks in the environment of the user. It works like a compass which is fixed at the bottom of the HMD. None of these concepts was elaborately tested with users.

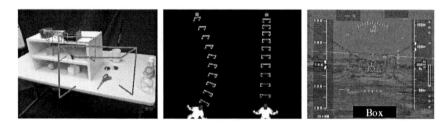

Figure 2.4.: Related Meta Visualizations: a)+b) The Tunnel concept to guide users wearing a tracked HMD (courtesy of [14]) c) Flight Tunnel in an airplane cockpit (courtesy of [77]).

Another metaphor is a *flight tunnel*, like it is used to guide pilots during their landing approach [77] (see Fig. 2.4 c). Biocca et al. [14] have used this flight tunnel metaphor to direct a user's view to the object in question. They used the tunnel to visibly link the head-centered coordinate space directly to an object-centered coordinate space (see Fig. 2.4 a+b). The tunnel is constructed by aligning elements on a Bezier curve between the HMD and the object to pick. They have proven in a basic experiment that the tunnel is better than having no Meta Visualization. However, in their scenario they did not require exact navigation, as we do in our scenarios. Within this work we adapt and iteratively improve the tunnel metaphor. In particular we had to find a solution to guide the user in tighter environments, as it is the case in the aisles of a warehouse.

Aisle Finding - Coarse Navigation

The coarse navigation guides the worker to the right shelf/aisle. We explored various visualizations to support this task in our first picking experiment [123]. We had solutions which were not applicable to real scenarios. Furthermore visualizations to support this coarse navigation have to be developed with the facts in mind that the worker makes large and fast physical movements when "running" from one shelf to the next, which means workers have to be prevented from stumbling, etc. Due to the fact that most logistics experts agreed that the number of shelves is generally manageable and that we did not have a huge warehouse in which to try it out, we did not focus on an Augmented Reality based solution for this navigation. Most of the existing industrial solutions focus on improvement of the Box Finding and not Aisle Finding. Finally we used convenient textual descriptions for the coarse navigation.

Box Finding - Fine Navigation

Much more critical is the fine navigation to one of the large number of boxes to pick from. This is where the worker probably can have the most benefit of Augmented Reality, as Augmented Reality shall make it possible to clearly highlight the box to pick from. The orange arrow in Fig. 2.3 shows an instance of a fine navigation. We soon realized that this is more complicated than we initially expected to highlight a box to pick from with Augmented Reality. Test participants in our first tests quite often picked out of the wrong box. The evolution of the fine navigation was a topic of optimization of the user evaluations described in Chap. 5. However, the fine navigation always has to be supported by a meta navigation. Both visualizations have to work together, which again can cause trouble.

Improving AR User Interfaces

Specific issues when evaluating Augmented Reality user interfaces. A report on lessons learned from several iterations of user-based evaluation in industrial environments. A new usability engineering strategy is proposed.

Each time a new technology comes along, new designers make the same horrible mistakes as their predecessors. Technologists are not noted for learning from the errors of the past. They look forward, not behind, so they repeat the same problems over and over again... The most egregious failures always come from the developers of the most recent technologies.

DONALD A. NORMAN, THE DESIGN OF EVERYDAY THINGS

Whether someone is going to use a technological product or not, depends mainly on the quality of the product. Aside from the traditional and fundamental quality criteria (e.g. performance and reliability), the key issues in the development of new and future technological products are utility and usability [74]. Utility describes the usefulness of a product and whether it meets the requirements of the user, whereas usability describes how usable a product is. Both issues are equally important [98]: It matters little that something is easy to use, if it is not what the user needs and it is also problematic if the system can hypothetically do something, but the user can not make it happen because the user interface is too difficult. Shneiderman et al. [132] even have the vision of having usability tags on user interface products, stating the effort one has to invest to use the product, similarly to refrigerators, which have tags to state their energy efficiency. This is not erroneous, as the business case for usability is strong and has been made repeatedly [80, 103, 12, 132]. However, sometimes marketing, a manufacturer's monopoly, or traditional reasons have larger influence on the commercial success of a product, than the product's quality has.

The problem is that designing highly usable and useful AR interfaces is fraught with problems. The largest challenge is that we work at the limit of what is known or common practice in the design of user interfaces [131]. That means Augmented Reality has a huge design space which allows the development of an immeasurable number of solutions, but at the same time presents an even bigger number of possibilities to fail with these solutions. Hence, AR user interfaces are mainly developed with a trial-and-error method. Furthermore Augmented Reality is still a technology in search of applications. There are only a very limited number of applications used in industrial settings from which we could get input for designing other successful applications. If there are existing solutions, we cannot simply transfer them to our settings, due to the diversity of the hardware, the environment and the users. Additionally, the number of AR user interface experts is limited, which is related to a lack of existing design and evaluation guidelines. Experienced users are also rare. All these factors already make it quite complicated to design a usable interface. It should be added that if a technology does not yet exist, the users can hardly imagine and express what benefits they could have from the technology and what undreamed solutions would be possible with it.

The only way to improve AR user interfaces, is to make intensive use of usability/utility testing. For classical 2D user interfaces there exist established usability/utility testing methods and it has been shown that such methods did not just speed up many projects, but also produced dramatic cost savings [121, 130, 31]. However, due to the just discussed problems of a huge, but unexplored design space and the limited knowledge on AR user interfaces and its evaluation methods, we have to find new and cost-efficient methods to evaluate and improve AR user interfaces. Shneiderman et al. [132] state that it is an exciting and provocative time in usability evaluation and practitioners should look at current procedures, and perhaps make some revolutionary changes.

To this end, the author reflects the experiences and lessons learned during several (Augmented Reality) user studies he executed (including the ones described in Chap. 5) on the traditional methods used to assess and improve user interfaces. For that purpose we give an introduction in to user interface design (Sec. 3.1), and therefore explain the user centered design process including usability and utility. In the next step (Sec. 3.2) we give established conceptualities on user studies and explain four important different types of experiments. After this mainly theoretic introduction, we explain practical solutions to problems occurring when trying to evaluate Augmented Reality user interfaces (Sec. 3.3).

User interface evaluations should never be executed in isolation, but rather should have a superior motivation/goal. As basically no mature Augmented Reality systems exist, the main goal of evaluating those systems should be to impact redesign and not to produce artificial problems to report about. For that reason AR evaluations should be executed within a utility/engineering process, but traditional suggested processes do not address the specific problems of immaturity and complexity to realize AR systems. To this end we propose an adaption of classical usability processes (Sec. 3.4) which takes the practical problems into account that the AR systems tested are mostly low-fidelity and that we do not have experts available either for AR user interface design or for the execution of usability studies.

Finally we present a short discussion on the acceptability of new technologies (Sec. 3.5), which is particularly important when trying to bring technologies like Augmented Reality to industrial workplaces.

3.1. Introduction to User Interface Design

This section gives an introduction on how to achieve utility and usability. The overall concept for this is a commitment to a user-centered design, which will be presented firstly. Then the terms utility and usability will be defined and explained with focus on Augmented Reality. This is followed by the rare few existing guidelines on how to design useful and usable AR user interfaces. The end of this section starts a discussion on the problems of evaluating AR user interfaces and on how to integrate the evaluations in the usability/utility engineering process. This discussion will be continued over the whole chapter.

3.1.1. User-Centered User Interface Design

The foundation for the development of usable and useful Human Computer Interfaces (HCI) – the user interface – is the commitment to a user centered-design by developing the system from the perspective of the user, rather than from a system view [101].

3D user interfaces are quite often new technologies in search of applications. Such a process unquestionably drives technological advancements. Yet, on the other hand, often forgets the user [33]. Getting the user involved in interaction development is now widely recognized as a key to improve usability of the interface [62].

For that reason, a distinction has to be made between the development of the *behavioral component* (how does the interaction work?) and the *constructional component* (how is the interaction implemented?) [62, 46]. Interactive systems have not always been designed and developed following this distinction, since what is best for the user is rarely easiest for the programmer. It thus requires a "team of three" to generate highly usable interfaces: a problem domain expert (user), a user interface software developer, and a user interface designer [62]. These should not just be roles but different persons, as each of the three has to advocate their goals.

To achieve a user-centered design we have to aim for a high degree of utility (Sec. 3.1.2) and usability (Sec. 3.1.3). A first step to achieve this is the application of guidelines (Sec. 3.1.4), which is problematic, as specific guidelines rarely exist. That is why we have to make intensive use of evaluations, which is also problematic for AR systems (Sec. 3.1.5). For that reason we discuss the problems of evaluating and improving AR user interfaces (in Sec. 3.2, Sec. 3.3 and Sec. 3.4).

3.1.2. Utility

The utility of a system predicates how useful it is and how well it fulfills the user's needs. We cannot apply simple task analysis to figure out what the user of an AR system needs. Because of the limited distribution of AR systems, users are rarely familiar with AR and thus cannot express or even imagine what this technology could make possible for them or what new needs it could fulfill. This is crucial for the success of an AR system as if we cannot find and show new beneficial possibilities of an AR system over a conventional method, we do not need to apply the expensive AR technology to that specific task.

However, the design of Augmented Reality user interfaces often focuses only on the usability of an application and forgets its utility. The requirements (or user-needs) elicitation is the process by which analyses determine the problems and needs of customers, so that system development personnel can construct a system that actually resolves those problems and addresses customer needs. This elicitation to increase the utility is an iterative process [59, 60, 3]. It might include state of the art reviews, focus groups and design workshops, scenarios, storytelling, personas, interviews and field observations [122]. A review by Anastassova et. al. [3] shows the lack of a structured methodology to analyze user needs in the area of Augmented Reality system development. The user needs analysis is a challenge for HCI methodology, because Augmented Reality is a technology in search of applications [4]. Thus users are required to express their "undreamed of" requirements [118], in particular during the early design stage.

User evaluations of mockups and prototypes are a powerful tool to elicit user needs [3]. Prototypes tend to facilitate the discussions about concrete functionalities [138] but their potential depends on the fidelity of the prototype [3]. Snyder [135] has shown that users respond more openly to low-fidelity prototypes than to high-fidelity ones. However, it is risky to evaluate low-fidelity prototypes as users tend to concentrate on usability problems instead of the generation of requirements. The test with the high-fidelity prototype leads to a much richer collection of information and, in addition, is more useful for redesign [4]. Even though the discussion about utility in this chapter is brief, it is equally important as usability.

3.1.3. Usability

Before we can start a discussion on how to achieve a high degree of usability in Augmented Reality interfaces, we have to make clear what we mean by usability. It covers more than just ease of use.

The definition everyone can comprehend immediately from bad experiences with user interfaces, comes from Tom Carey (cited in [62]): "If your computer were a person, how long 'till you punch it in the nose?". Shneiderman et. al [132] cite the ISO 9241 Standard (Ergonomics of Human-System Interaction) usability goals: effectiveness, efficiency and satisfaction. However, they give a further definition, which they think leads more directly to a practical evaluation of the term usability:

- *Time to learn* the actions relevant to a set of tasks.

- *Speed of performance* to carry out benchmark tasks.

- *Rate of errors* How many and what kind of errors do users make in carrying out the benchmark tasks?

- *Retention over time* How well do users maintain their knowledge after a period of time?

- *Subjective satisfaction* How much did users like using various aspects of the interface?

Bowman et. al [17] proclaim that "usability encompasses everything about an artifact and everything that affects the person's use of the artefact". Following this definition, Shneiderman's measures do not take into account the physical and psychological demands of 3D user interface (e.g. 3D user interfaces often use fancy and bulky display hardware and produce cybersickness). Altogether such interfaces can produce fatigue and strain. To this end we add the following usability criteria:

- *Physical Strain* Does the user interface produce physical strain, when used for a usual amount of time? Is the user interface ergonomic?

- *Psychological Strain* Does the user interface produce psychological strain, when used for a usual amount of time?

There are some other factors, which directly affect the perceived usability and the willingness of tolerating usability errors and thus support the willingness to learn the interface: users prefer more aesthetic or beautiful products [143, 79, 100]. Other aspects which motivate to use a product are *Hedonic Quality* and *User Experience* [54]. Hedonic Quality is a combination of how much users can identify themselves with a product and how much this product stimulates the users' creativity and self development [54]. User Experience defines the degree to which a user has a positive feeling after using a product [38]. For that reason we extend the list of usability criteria with the following, thereby using the term pragmatic as something which is related to utility.

- *(Non-Pragmatic) Attractivity* Does it look good?

However, one important aspect to achieve a high degree of usability (and utility) is notably absent from current AR efforts – the availability of established and recognized design and evaluation guidelines which specifically address usability [43]. That is why the rest of this chapter contains a discussion about the design and evaluation guidelines for Augmented Reality.

3.1.4. Design Guidelines

A starting point for a user-centered design is to rely on common techniques, in particular, to follow guidelines. We give a short introduction into general design guidelines and mention the rare few existing guidelines for AR user interface design.

General Design Guidelines

The most universal guidelines are Norman's "Principles of Good Practice"' [102]: a) *ensure a high degree of visibility*, by allowing the user to work out the current state of the system and the range of actions possible, b) *Provide Feedback* by giving continuously clear information about the results of actions, c) *present a good conceptual model* by allowing the user to build up a true picture of the way the system holds together, the relationships between its different parts and how to move from one state to the next, and d) *offer good mappings* by aiming for clear and natural relationships between actions the user performs and the results they achieve.

Shneiderman et. al. [132] are more specific in framing their "eight golden rules of interface design": 1) *strive for consistency* in actions, terminology, layout, and design, 2) *cater to universal usability* - e.g support novices and experts, 3) *give informative feedback*, 4) *design dialogs to yield closure* - e.g showing a clear start and end point, 5) *prevent errors* - e.g. avoid/detect wrong user input, 6) *permit easy reversal of actions*, 7) *support internal locus of control* - the user should always have the sense of being in charge of the interface, and 8) *reduce short term memory load*.

Further design principles, style guides, and standards can be found in [96, 137, 62, 132]. Hix et. al. [62] state that one should not take the fact of having followed design guidelines and standards as an excuse not to evaluate the final user interface. However, these guidelines are a good starting point and definitely important for the general design and evaluation of (AR) user interfaces, even though they do not address the specific problems of AR user interfaces.

AR Specific Usability Design Guidelines

The general design guidelines for 2D desktop based user interfaces have emerged over the last decades. Among other reasons the relatively fast changes in the infrastructure and capabilities of hardware and the general diversity of hardware and applications mean no dedicated design guidelines have yet been established for 3D interfaces. We are therefore still limited to an exploratory trial and error approach when designing such interfaces [117].

Dünser et al. [33] make suggestions on how to apply the general design guidelines to the design of AR interfaces. Besides keeping Normans' "Principles of Good Practice" in mind, he suggests the following generic principles: *Reducing cognitive overload, Learnability, User satisfaction,* and *Responsiveness and feedback*. Furthermore he suggests the more

AR related topics: *Affordance* – give an inherent connection between the user interface element and its functional and physical properties (e.g. tangible user interfaces). *Low physical effort* - prevent users from unnecessary movements, and use light weight and ergonomic HMDs, *Flexibility in use* – address different user preferences and abilities, by the integration of different I/O devices or interaction techniques, and *Error tolerance* – e.g. compensate for instability by integrating intelligent/hybrid tracking. By doing so he gives evidence to the validity of the general guidelines for AR interfaces. This goes along with the statement by Shneiderman et. al. [132], that the general principles must be interpreted, refined, and extended for each environment. They agree with Bowman et al. [17], that designers should be able to rely on the general guidelines, but need to come up with specific solutions for their individual problems. Establishing general guidelines for AR user interfaces is problematic due to various kinds of displays, environments, interaction techniques and platforms [33].

AR Specific Utility Design Guidelines

Regenbrecht at al. [113] state guidelines for the design of useful industrial AR applications. These guidelines are the result of reviewing 10 industrial projects: *Data Integration* – the early integration of real world data is crucial for the successful deployment of the system. *Simplicity* – strive for a simple and elegant solution not for the one with the most originality and novelty. *Added Value* – even if the figures are "educated guesses" look for a return on investment appraisal. *Acceptance* – the project team should find a supporter of the project within the company.

3.1.5. Problems of User Interfaces Evaluation and Usability Engineering for AR

This section gives an introduction to the problem of evaluating Augmented Reality user interfaces. We discuss problems of expert- and user-based evaluations and raise the question of the best way to evaluate AR user interfaces.

Informal tests (e.g. with colleagues) can provide some useful feedback, but more formal expert reviews have proven to be far more effective [137, 99] and can help to identify usability problems early in the design stage. A common technique for expert based evaluations is a *Heuristic Evaluation*, where experts test the user interface for compliance with recognized usability principles such as the just discussed guidelines (see sec. 3.1.4) [96]. Results of a heuristic evaluation can be the basis for user evaluations carried out in the later design process [44]. However, this technique as applied to AR interfaces has a problem: the lack of comprehensive guidelines. Furthermore Augmented Reality-HCI-experts rarely exist and those that do are mostly not easily available [132]. It makes an enormous difference if the experts are familiar with the rules and are able to interpret them. Other forms of expert evaluations are *Cognitive Walk Throughs*, *Consistency Inspections*, and *Formal Usability Inspections* also known as a courtroom-style meeting. Reviews should be placed in a situation as similiar as possible to the one that the intended users will experience, in an as realistic as possible work environment, complete with noise and distraction [132]. The danger with expert reviews is that experts may not have an adequte understanding of the task domain or user communities. Furthermore experts come in many flavors, and conflicting advice can further confuse the situation – cynics even say "for every Ph.D. there is an equal or opposite Ph.D." [132].

Usability testing and controlled experiments remain the backbone of evaluation [26]. They are much more effective in finding usability problems that novice users encounter than heuristic/expert based evaluations [42]. In fact, laboratory studies are often the only time researchers and developers spend observing users using their user interfaces [110]. However, if user evaluations are conducted out of incorrect motivation or if empirical methods are not properly applied, the reported results and findings are of limited value or can even be misleading [32].

Today, complex systems that are hard to test with simply controlled experiments abound [105]. This means that there seems to be a lack of evaluation methods for AR interfaces. All methods have their distinct advantages and shortcomings. They supplement each other since no single method can identify all problems [62]. This raises what is probably the most important open question when developing highly usable AR system [43, 117, 45, 33]:

> *What guidelines shall be applied and what kind of existing or new evaluation method shall be applied to measure and improve the usability of an Augmented Reality interface?*

This chapter points out ways to get closer to an answer to this question. A first step is to remember the quote by D.A. Norman in the introduction of this chapter. We have to look back to see what worked in the past and reflect on the way we used to evaluate and redesign our system (see Chap. 5) and thus conclude and adapt the established techniques to the evaluation of AR interfaces. To this end we have to adapt the traditional *Usability Engineering* processes to meet the requirements of developing user interfaces within the largely unexplored AR design space. In the past, highly usable 2D desktop based interactive systems were built by applying *Usability Engineering*. Usability engineering is a cost-effective, user-centered process that ensures a high level of effectiveness, efficency, and safety in complex interactive systems [62]. According to Hix et al. [64], usability engineering includes both design and evaluations with users; it is not typically extensive hypothesis-testing-based experimentation, but instead is a structured, iterative user-centered design and evaluation applied during all phases of the interactive system development life cycle. It basically consists of the following phases: user task analysis, user class analysis, design of the user interaction, rapid prototyping, expert evaluation, user-centered evaluation, and iterative redesign based on evaluation results [62, 61]. This definition of usability engineering seems to be true and approved for the development of traditional and established 2D user interfaces, but has to be adapted for the development of AR user interfaces.

Even if existing for more than 40 years, the main problem when we apply Augmented Reality to a new area in a new setup is that we quite often have to find new solutions for fundamental problems, while at the same time trying to develop a comprehensive system. This requires us not seldomly to invent interaction techniques from scratch, or see how established techniques perform in new environments. This again requires extensive experimentation to figure out the basics of human interaction with Augmented Reality systems in such situations. This means we have to adapt the established usability engineering approaches, which basically focus on fast and formative development, to the development of highly complex and unexplored AR interfaces.

To this end, we propose an adaption of established usability processes in Sec. 3.4. To provide the rationale for this adaption we present an overview of general user study designs in Sec. 3.2 and discuss the particular problems of setting up, running and analyzing user studies for Augmented Reality interfaces in Sec. 3.3.

3.2. User Studies & Structured Experiments

This section gives an overview of the basics of empirical experiments and makes a fundamental classification of different types of experiments: formative, summative, human factors and case studies.

3.2.1. Fundamentals of Empirical Experiments

The standards used in empirical research methods make a difference between scientifically valid data and ordinary observations [155]. Zimbardo [155] states the following requirements:

- *Objectivity* One has to follow strict rules to not color the results (e.g. by setting up the actual hypothesis to prove after the experiment). Furthermore the whole experiment and its analysis has to be strictly documented and reported, so that others can reproduce, understand and comment on the experiment.

- *Reliability* Experimental methods have to be standardized and sources of unwanted variations have to be controlled, by for example taking sufficient observations, to prevent individual atypical effects from distorting the results.

- *Validity* Does the experiment really measure what it should measure? E.g. does the test measure the real I.Q or does it provide an erroneous measurement to be relaxed under test conditions.

There exists a standard terminology for empirical experimentation:

- *(Test) Participant* or *User* is the term used in this work to refer to someone taking part in an experiment. We do not use the term *subject* anymore to ensure that people feel that they are treated with respect.

- *Independent Variables* are all those variables which are systematically varied during the experiment. E.g. a variable can have three values blue, red, or green. It is then said to have three *levels*.

- *Dependent Variables* are all those variables which are observed and measured during the experiment. The goal of an experiment is to find a dependency between the dependent variables and the levels of the independent variables. Typical dependent variables are task completion time and errors made.

- *Confounding Factors* are all those factors which have an unwanted influence on the result of the experiment. Confounding factors have to be removed (if possible) or at least have to be kept constant. A confounding factor can be for instance the experience level of the user (novice, expert), learning effects or exhaustion.

- *Evaluator* is the role of the person who is executing the experiment. An evaluator can have several assistants.

- *Hypothesis* is the expected result of an experiment. It often happens in our young area of 3DUI research that hypotheses are setup afterwards to match the statistical analysis. However, it is strictly required to setup the hypotheses (if there are any) before the experiment, otherwise the criterion of objectivity (which was discussed above) is not met.

- *In-between Subject Design* is typically used when many subjects are available and the execution under the individual experimental conditions takes some time. It means that each user is only tested under one level (or treatment). It is in contrast to with-in subject design.

- *With-in Subject Design* is the typical experimental design when only a few users are tested and the experiments are relatively short, but the independent variable has several levels. It means that every user is tested on more than one condition. To compensate for learning effects, the order in which the different levels are tested have to be permuted for each subject. It has a greater statistical strength than the in-between subject design as it reduces the error which can be introduced by individual differences.

There exist three types of coherences which can be observed when executing an experiment: 1) *causality*: if one occurrence causes the other, 2) *correlation*: two occurrences regularly occur together, without one causing the other and 3) *coincidence*: two occurrences accidentally but not regularly occur together. The goal of a (psychological) experiment is to find cause and effect relationships – causal dependences, where an effect (change in the dependent variable) only occurs because of a change of the independent variable.

3.2.2. Types of Experiments

The classification into different types of experiments has to be understood by every designer of experiments. In retrospect of this work, this is one of the most important classifications, as the type fundamentally influences the experimental design. The distinction is not in its formality, but rather in the goal of each approach. Thus it is important that members of the development team, and especially managers, understand this difference. Otherwise the results of a formative approach may be discounted as being, for example, too informal, not scientifically rigorous, or not statistically significant [62]. While academics were developing controlled experiments to test hypotheses and support theories, practitioners developed usability-testing methods to rapidly refine user interfaces [132]. The first two definitions where coined by Scriven [128].

- *Formative Evaluation* is typically conducted during the development or improvement of a program or product [129]. It is done in order to improve the interface as part of the iterative design process [98]. It does not look for statistical significance nor claims to be scientifically rigorous. It rather supports redesign, followed by another evaluation. It produces quantitative data against which developers can compare the established usability specifications and also produces qualitative data that can be used to help determine what changes to make to the interaction design, in order to improve its usability [62].

- *Summative/Comparative Evaluation* compares a finished design or product to a concurrent/alternative product or it simply has to prove if the product meets some specifications. It is executed for the benefit of some external audience or decision makers [129].

- *Human Factors Evaluation* is close to a psychological experiment. Shneiderman et. al. [132] state: "It enhances the classical psychological methods to deal with complex cognitive tasks of human performance with information and computer systems. The main goal of human factors experimentation is to find out about the basics of the human factors in the interaction with machines. It therefore makes use of intensive hypothesis testing and intensive data analysis. The reductionist approach required for controlled experimentation yields narrow but reliable results to this field and thus each small experiment result acts like a tile in a mosaic of human performance with computer-based informatiomn systems". This is in particular interesting for the optimizations on tasks which are executed thousands or millions of times a day (which happens in industrial scenarios). Those tasks can already benefit from interfaces which provide an improvement of seconds. However, there is a huge difference between human factors experiments in mature sciences compared to AR. In AR we still have so many confounding factors influencing our experiments that it is not this easy to conduct human factors experiments.

Human Factors experiments do not generate results in a time frame that meets the needs of the fast cyclical iterative process in product development. The iterative process of formative usability evaluation is the main area of usability evaluation, whereas summative evaluation, in practice, is hardly applied [62]. However, there is a crucial point when trying to develop and evaluate usable AR systems. We cannot simply apply the iterative formative approach, as our knowledge about human factors in the AR interface design space is very far from being comprehensive. During the formative process, we continuously encounter fundamental human factors problems. Every time such a problem shows up, we actually should solve it by executing a reasonable human factors experiment, using of a huge number of participants and intensive hypothesis testing. This seems not to be feasible in a process that has to be efficient to a certain degree, as every project (it does not matter whether it is in academia or industry) has to be finished within a limited time frame. The solution proposed in this work is an efficient integration of human factors experiments into the formative evaluation process. This is achieved by executing low-budget human factors experiments with a limited number of users, and thus finding good and working solutions. Results from these low-budget human factors experiments have to be reported to the human factors research community, to make people find the general solution with high reliability, as will be discussed in Sec. 3.4. Finally we extend the general classification of experiments by the following category.

- *Case Studies & Ethnographic Approaches* Scientific evaluation based on controlled experiments has its weaknesses: it may be difficult or too expensive to find adequate test users, and laboratory conditions may distort the situation so much that the conclusions have no application [132]. Because of that laboratory experiments have to be supplemented with case studies, reporting on users in their natural environments, pursuing real taks, over a longer period of time [110, 109]. The weakness of such case studies is that they are very time-consuming and the results may not be universally valid. However, Shneiderman et al. [131] even promote the execution of long term in-depth cases studies and making use of ethnographic methods. Ethnography originally comes from social/cultural sciences and describes the method to observe a foreign culture by temporarily being part of the culture and their daily life. The main idea is to describe the culture from their own perspective. Hence, applying ethnographic methods to the evaluation of user interfaces means that the evaluator becomes a co-worker of the user rather than a withdrawn neutral observer. The evaluator and user explore the user interface together in the user's natural environment while performing real tasks.

Thus far the theories of empirical evaluation have been discussed; the practical aspects of evaluating AR user interfaces follow.

3.3. Running User- and Expert-based Studies for AR

The actual design of an evaluation is influenced by several aspects. It is most important to have an objective as to why to execute the experiment at all. This is the base for a detailed evaluation plan [62]. The real goal of usability studies is to impact design and not to produce problems, which are potentially not real [153]. According to Shneiderman et al. [132], the evaluation plan is determined by the stage of design, the novelty of the project, the number of expected users, the criticality of the interface, the cost of the product and finances allocated for testing and, because our own empirical knowledge is quite important, the experience of the design team. These factors directly influence the choice of the experimental type (as just discussed in Sec. 3.2.2). This is either a formative, summative, or human factors evaluation, or rather a case study.

Shneiderman et al. [132] argue that "early research in human-computer interaction was done largely by introspection and intuition, but this approach suffered from lack of validity, generality, and precision. The techniques of psychologically oriented, controlled experimentation can lead to a deeper understanding of the fundamental principles of human interaction with computers. This includes proposing a lucid and testable hypothesis, the consideration of an appropriate research methodology, the execution of the experiment, and collecting and analyzing the results". On the other hand, many problems are often discovered by accident [72] before, during and after the evaluation itself [67]. So we have to take care that we do not waste our resources, as the execution of formal experiments can be quite expensive and time consuming.

This section discusses aspects to be taken into account when performing user studies with Augmented Reality prototypes. It starts with the setup (Sec. 3.3.1), continues with what to be considered when running the session (Sec. 3.3.2), and ends with the analysis of the data (Sec. 3.3.3).

3.3.1. How to Set Up

This section presents some aspects to be considered when setting up an experiment to evaluate Augmented Reality user interfaces. This includes the usability lab, practical problems to consider, and a discussion on the test participants and the evaluators.

Usability Lab To perform good user studies, it basically requires a room with a closed door and a scratchpad for noting down observations. This room should have a sign on the door marking it as a usability lab, as it makes test participants when entering the room feel that they are making an important contribution. The sign furthermore makes an organization's commitment to usability clear to employees, customers, and users [62, 132]. Having special equipment and tools for data logging and recordings of observations can be helpful to handle the vast quantities of data and helps to focus on observing the participant rather than on hectic note taking. Performing user studies with Augmented Reality systems, requires either bringing the elaborate equipment to the usability lab or to have a copy of the expensive hardware permanently available. Observing the user via a classical one-way mirror and a simple video camera setup is often not possible as users usually move around rather than sit in front of a desktop application. This can require inventive observation positions or adventurous camera setups (e.g. multi, mobile, head-mounted, wide angle) setups, which moreover have to be synchronized with each other and the AR visualization.

Practical Problems of the AR Setup in an Evaluation The evaluation of Augmented Reality is associated with a lot of practical problems, as AR requires user/object tracking and quite often immersive but inconvenient display devices. Typically the tracking setup needs to be calibrated before each session, because it quite often happens that someone has moved measured objects/sensor components or a sensitive marker-setup has been bent. The best calibration has to be used for the experiment, as registration errors can confound the experiment, influencing usability and acceptability [6]. In the case of industrial setups, the setup sometimes has to be redone for each session, because the environment is used by routine business at other times [58]. As it is the main idea of Augmented Reality to augment the real environment users are usually interacting heavily with the real environment during the experiment. Thus users may stand rather than sit, and they may have to move around in a wide space using whole body movements. This aspect gets more pronounced the more the user is isolated from the real world due to tunnel-vision HMDs, high latency, or cluttering visualizations. In consequence, the evaluator(s) need to take care that users will not bump into walls or other physical objects, trip over cables, or move outside the range of the tracking device [16]. It is difficult for the evaluator to see what the user sees [16]. This can be a significant drawback for the observer. In mobile optical see-through HMD setups, where it is impossible to see exactly what the user sees, it is useful to show at least the augmentation on a computer screen, which is, for example, mounted on the back of the user. This does not allow the observer to see the registered augmentation, and certainly does not provide information on the quality of the registration, but he at least can see what kind of virtual objects are currently in the user's field-of-view. This is in most cases enough to see why the user struggles.

Unexplored Design Space As already mentioned, Augmented Reality still has a huge unexplored design space; minimal changes in the design can lead to completely different results. For example in their first experiment [142] Tönnis et al. could not find significant evidence for an AR visualization to be better than a 2D birds eye visualization to direct the car drivers' attention to upcoming danger. However, by improving the AR visualization and the setup, they proved AR to be better in a succeeding experiment [141]. That means, when performing statistical experiments, it is often difficult to know which factors have a potential impact on the results. Besides the primary independent variable, a large number of other potential factors could confound the experiment. One approach is to try to vary as many of these potentially important factors as possible during a single experiment [16]. However, this actually leads to complex and extensive experimental setups. To this end we have to find a cost-effective way to reduce the design space before executing expensive experiments.

Test Participants & Sample Size First of all, the test participants should be treated with respect and thus they should not be called subjects, as it can make them feel like rats in animal testing [62]. There is always a discussion about the number of test users needed. The formative evaluation process does not focus on a large number of participants; it rather intends to extract as much information as possible from every participant [25, 152, 62]. A widely used assumption is that 3-5 randomly chosen users are the optimum number of users for one cycle of formative evaluation, because they will help to identify about three-quarters of the usability problems [96, 95]. This method is also known as "discount usability testing". Other research showed that 5 users only identified 35% to 55% of all problems [136, 36]. However, Lindgaard et. al [84] showed in 9 tests that careful participant recruitment and investing in wide task coverage is more fruitful than increasing the number of users. Most of this research was done in the area of web usability. Our experience is when testing Augmented Reality interfaces in the early design stage (which is where most AR interfaces are), a few users are enough to identify most usability problems. The users typically focus on the same embarrassing usability problems and therefore will not identify other ones. Thus if the evaluator realizes that the feedback of the test participants starts repeating itself, the iteration of experiments should simply be stopped, the problem fixed and a new cycle should be started.

The feedback during an evaluation strongly depends on the type of test user. An evaluation with HCI-experts cannot serve as a substitute for an evaluation with representative test users [62]. Thus people from both groups have to be involved. The problem is that very few experts for virtual environments [16] or even augmented environments exist.

Such experts will offer suggestions for fixing usability problems, unlike the representative user, who typically tends to find a problem but can not offer suggestions for resolving it [62]. The representative user will give domain specific input of high value, which is critical for increasing the utility of the system. We realized that another type of test user is quite important: the inexperienced random user, who is pulled from the hallway into the lab by chance. Such users do not focus on fixing usability problems nor do the try to give domain specific input, they just explore the system from a neutral point of view. After having done tests with two users of each of the three groups, one has not exactly tested a perfect subsample, but in most cases one obtained enough input for the main aspects.

Things are different when executing a comparative or an exhaustive human factors experiment. Those experiments mainly look for reasonings on the basis of statistical hypothesis testing. Most labs seem to have a magic number of test users; everything above 12 is typically accepted. Theoretically, the number of users can be calculated, if one takes the effect size into account, as described in any statistics book [15]. If the effect between the different test conditions is large, it is enough to collect quantitative data from a few test participants. In that case the statistical tests easily show significant differences, even if the sample was not perfect and the data are badly distributed. If the effect size is small, more users are needed. However, the decision on the number of users has to be made taking into account economical factors. If it is important to know that one technique is a few percent better than another, because small productivity improvements would translate into large savings, investments into more test users should be made.

Another economic policy requires performing comparative discount usability tests until one is "almost sure" about the results of an experiment. In that case, the exhaustive experiment with "enough" test users is only executed to collect enough quantitative data to get significant statistics for the stakeholder or scientific publications.

Evaluator   The team of evaluators plays a critical role in the evaluation. Due to the complexity of AR-setups, most experiments require more than one evaluator [16]. In our experiments, we realized that it can be quite effective to have a dedicated evaluator, who is simply there for observing the participant and does not have to take care of the procedure and handling test participant. This observer can fully concentrate on the behavior of the participant, in order to ask specific questions in a later interview. A study of evaluations in industry found that different teams of evaluators identified different usability problems [92]. Further, attention has to be paid to the evaluator's knowledge and experience, which must not be underestimated. Heuristic evaluation and usability testing draw much of their strength from the skilled evaluators who execute them [73]. Nielsen [97] showed that persons with knowledge on usability evaluations found about two to three times as many usability problems in an interface as novices did.

Tasks Most of the usability criteria (as discussed in Sec. 3.1.3) cannot be directly measured. We thus have to find aspects of usability that can be measured. This means, that we have to find measurable and representative tasks that users are expected to perform often [62]. At the same time we must not forget the other tasks which add breadth to the evaluation. The choice of such measures not only fleshes out what usability means, it also raises the question if the measured data is a valid indicator of usability [66]. Even when only performing semi-formal walkthroughs, it is quite important to let the users really perform the task (i.e. picking the item), instead of just letting them do it verbally ("And now I would pick the item from that box"). Only simulating the task quite often overlooks important aspects.

Planning the Session & Pilot Testing A test session needs to be planned in detail, as everything which goes wrong during an experiment can confound the results. Furthermore, every minute one makes the user wait, due to preparation or bug fixing, is lost time in which one could have actually tested the (expensive) user. To compensate for that, the evaluator must perform some pilot testing to ensure that all parts of the experiment are ready. Pilot testing requires a very small amount of time compared to all other effort [62].

3.3.2. How to run the Session

This section describes some aspects to be considered during the test session. In particular it addresses the problems of introducing the user to the test setup and to Augmented Reality. This discussion is continued with a discourse on whether and how to actively observe and interrupt users, when they are struggling with the new technology. Furthermore, we suggest some observation methods and techniques.

Introduction to the Session Many researchers agree to give uniform instructions, (in the best case by some objective medium) and to treat everybody in the same way during the whole experiment. This way, all participants start with the same level of knowledge about the system and the tasks they perform [62]. This helps to ensure objectivity [155]. We proceeded like that in our first experiment (see Sec. 5.1), using carefully prepared Powerpoint slides. The experiment failed dramatically. The reason for that was that trying out Augmented Reality (e.g. for novices) is so far removed from what most people know, that there can be a high fascination effect, accompanied by a huge learning curve. Some people need a long basic introduction into AR. They do not comprehend from the outset the concept of the spatial alignment of virtual and real information when they wear an optical see-through HMD. On the other hand, fast adapters to the technology can start almost immediately to use the system efficiently. For that reason, different people need completely different introductions. We finally replaced the uniform introductions with individual try-and-ask phases. The try-and-ask phase is like a cooperative exploration of the user interface. The participant is obliged to try and ask until the evaluator thinks the participant fully understands the interaction technique or system. This concept was successfully introduced in the second experiment (Sec. 5.2) and improved in the further ones. It reduced the measurable learning effects which confounded the results of our first experiment. Furthermore, this try-and-ask phase is already a first assessment of the user interface. The questions asked by the participants in order to understand the user interface can be good indicators of usability problems.

Directing the Session and How to Observe When performing quantitative data measurements, the evaluator should generally remain in the background and not interrupt the test participant [62, 16]. However, we found several important reasons for such interruptions.

In particular, when performing longer experiments with AR-setups, we observed that users struggled with a general problem (e.g. the HMD did not sit perfectly) or the user used a strange strategy (making acrobatic movements). In such cases, we interrupted the experiment, because such behaviour obviously has an influence on the measurement, which is stronger than the different treatment/level of the independent variable. In such situations, careful note taking is required to delete the wrongly measured performance. Sometimes, when a second participant was seen to discover (and struggle with) the same usability problem, we decided (against the established rules) to guide the participant, while noting the discovered usability problem.

The problem is how to observe the user and how to collect the important qualitative data (which is more important than quantitative data in the early design stage). The famous baseball player Yogi Berrad said, "you can observe a lot by watching." The problem is, you have to learn how to watch [102].

A common technique to get a lot of problems and solutions reported from the test participant is the *think-aloud method* [35]. The basic idea of this method is that users tell what they think when exploring the user interface or performing tasks. Jorgensen [75] states that one positive aspect of the think-aloud method is the experience of developers seeing users struggle with an interface which has been designed by them.

Crtitical Incident Taking [29] is another way to collect qualitative data. A critical incident is something that happens while a participant is working, that has a significant effect, either positive or negative, on task performance or user satisfaction, and thus on the usability of the interface.

In our later tests (Sec. 5.7), we introduced the *Muppet Show Balcony* as an observation method. By this, we basically mean that the observer literally sits somewhere above the area where the experiment is executed. Thus the observer is not noticed by the test participant, but can "critique" them. In allowing the evaluator to critique the participant, we mean that he is allowed to interrupt the participant in critical situations to ask questions or to guide the participant. This situation is similar to a one way mirror. The observer is still distant and unobtrusive, but when talking to each other, they can see into each other's eyes. This way they feel more welcome and not like lab rats behind a window. Furthermore the balcony allows taking notes without the participants realizing it. We report on this in Sec. 5.7.2.

Interview Observations made during the experiment can be followed up in a final interview. This interview should be semi-structured, i.e. a series of preplanned as well as open questions should be asked. However, it takes some experience to get valid and reliable answers. The interviewer could unknowingly force specific answers, or participants could feel awkward and thus just give positive answers [155], which would be a classical example of the evaluator effect [155].

Questionnaires Aside from the active involvement of the user, questionnaires are a generally accepted companion for usability tests. A widely used technique to generate a questionnaire is to base it on the semantic differential [134] (opposed adjectives), or a Likert Scale [83] (ranking of agreement-disagreement to an item). Ready questionnaires are, for example, the Systems Usability Scale [21], or AttrakDiff^{TM} [23]. It is always advisable to use such standardized questionnaires, as they are proved for validity. Selfmade questionnaires mostly lack such validity.

Measuring Strain We defined physical and psychological strain as additional usability criteria (Sec. 3.1). To this end, we will briefly present some tools to measure them. A cheap method to determine the mental workload is the NASA Task Load Index (TLX) [53]. It is based on a self assessment. This is problematic when only having a small sample size. Alternatively the "d2 attention strain test" [20] can be used to measure the concentration before and after the test. A fully objective approach, which however requires a lot more effort, is to make use of a Heart Rate Variability analysis [146]. For this, the user wears a heart-beat recorder. We made use of this in one of our experiments (compare Sec. 5.7).

3.3.3. How to Analyze and Proceed with the Results

The intermediate result of an experiment is typically a huge amount of data, which has to be analyzed and interpreted. The final result of the experiment should impact the redesign of the system. It provides a collection of insights, findings and lessons learned which should be discussed with the research community.

Plaisant [110] states that "statistically significant differences in time, error rate or satisfaction are obviously a plus, but observations recorded during the test become the basis for refinement or redesign, leading to better implementations, guidelines for designers and refinement of theories. These benefits are very appealing to researchers but much less for potential adopters of the technology who are left wondering what the performance would be with the improved interface and what would have happened if users had been trained longer. Including pratical summaries for practitioners in research papers is helpful. Reporting informally on usage and performance by developers can shed some light on the potential utility of the tool and performance of the trained users".

Quantitative Data Analysis & Hypothesis Testing   We have reasoned to not mainly focus on the statistical analysis when evaluating the results of an experiment, but we should not exclude the statistical analysis from it. Therefore we briefly explain some aspects of exploratory and inferential statistics, but our general advice is to consult an expert when planning to perform statistical analysis.

For that reason, we give only an example of how to perform a statistical analysis by explaining the way the analysis within this work was performed. This section is based on several statistics books [15, 139, 154, 111]. It is important to keep in mind that there is always not just one way of carrying this out, and for that reason the way statistical analysis is done has to be reported. This permits the reader to reproduce, understand and comment on the experiment and thus helps to ensure objectivity.

A data analysis is mainly performed to get answers for the exploratory research question or to prove or reject the hypothesis. This is mainly performed by statistical tests of significance as mathematics seem to be the most convincing argument in proving something. But we should not reject/accept hypotheses only because of statistically significant differences [15]. Anecdotal evidence or individual insights may be given too little emphasis because of the authorative influence on statistics[132]. Hence it is important to perform some kind of exploratory data analysis and to keep the qualitative observations in mind to decide whether a test on significance is advisable. Quite common techniques for this exploration are to generate *Histograms* (graph with tabulated frequencies), *Box Plots* (a graph which shows among other information: minimum, maximum, quartile and median, outliers) or *Bar Charts* showing mean values and standard deviations.

If the data exploration does not show peculiarities we can proceed with the statistical analysis. If we follow the strict mathematical way of hypothesis testing, we have to reject the antithesis H_0 of our actual hypothesis H_1 to be able to accept our actual hypothesis. H_1 is called the alternative hypothesis and states that different treatments produce different observations and H_0 states that there is no difference. The actual goal of these hypotheses tests is to draw conclusions from the test sample on the population. There are two possible errors doing this: 1) false positive – H_1 can be wrongly accepted: α-error or 2) false negative – H_0 was wrongly accepted: β-error. The commonly accepted

α error for this kind of scenarios is 5%. The β-error can be calculated from the variables of the experiment.

The actual method to test for significancy depends on the presented data. In the case of a normally distributed sample, parametric tests can be used, otherwise non parametric ones have to be used. If the sample is smaller than 30, typically the Kolmogorov-Smirnov test (KS-test) is used to test for normal distribution. In the scenarios of this thesis we mostly make use of small sample sizes and thus use a within-subject-design (Sec. 3.2). This means each user is tested under all different conditions which is called repeated measures (RM). Thus we can make use of tests which compare the values per user, rather than the values of the different treatments in general. Those "dependent" tests lead to more valid results in case of small samples and within-subject designs. The parametric tests we make use of in this work are repeated measure t-test (T–RM) and for more than two treatments repeated measure ANOVA using a general linear model (GLM–RM). The GLM-RM tests only for an existing difference. The actual difference has to be estimated by a post hoc analysis, in cases of this work the Fisher's Least Significant Difference test (PH–LSD) will be used. The nonparametric equivalents for the tests are Wilcoxon (Wil) for two treatments and Friedman (Frie) for more than two treatments. However these tests are less powerful than the parametric ones and give only a range for each treatment.

3.4. Usability / Utility Engineering Processes for AR

So far, this chapter has discussed the important aspects and the influencing factors of evaluating and improving Augmented Reality user interfaces. To sum up, the design and development of highly usable and useful Augmented Reality interfaces comes with many challenges. It is already complex to set up an Augmented Reality system in general, because it requires a perfect interaction and calibration of tracking systems, application software, and display hardware. Aside from the fact that we have not yet experienced a system in which these problems were solved perfectly, the actual challenge of the interface design actually starts afterwards. The design space is highly unexplored and one cannot refer to Augmented Reality experts, and design and evaluation guidelines. All these aspects together guarantee that the first system designs fail, and this is probably true for every new system being designed. Additionally, it already requires comprehensive expertise to set up, execute and analyze a classical empirical experiment. It becomes even more complicated when the usability and utility of AR interfaces have to be assessed and improved. Many factors can easily influence the performance of users with an AR system and thus extremely confound the results of an experiment.

To this end, we suggest an iterative and efficient usability/utility engineering process for AR systems which takes these problems into account. It aims to bridge the gap of lacking knowledge on AR design and evaluation guidelines and it integrates a circular evaluation loop into the classical process, which allows designers and evaluators to explore and conquer the design space stepwise and make them more and more experienced. However, before presenting our process we will briefly discuss existing approaches.

3.4.1. Usability/Utility Engineering Processes: State-of-the-Art

A typical problem occurs when designers and evaluators do not follow a carefully thought out evaluation strategy. They mostly forget the essential formative evaluation step. This may result in a situation where the expensive summative evaluation is essentially comparing "good apples" to "bad oranges"[63]. For example, we have developed a new interaction technique for AR and we want to show in an experiment that it is better than an established solution. We fail in most cases if we have not shaped our interaction technique sufficiently in formative evaluations. Conversely, consider if we have developed an interaction technique and improved it over several iterations. For some reason we want/ have to show that our interaction technique is better than an established solution. To this end, we develop an ad-hoc version of the "established solution". An experiment comparing both techniques will of course prove our new solution to be better.

To avoid such situations, we have to follow at least some ordered steps in an evaluation process. A classical evaluation approach for virtual environments is similar to the one described by Gabbard et al. [44]. It basically consists of the four sequentially ordered steps: 1) User Task Analysis, 2) Heuristic Expert Evaluation, 3) Formative Evaluation 4) Summative evaluation. It is not strictly a "straight process", but advises loops of the first three steps. In recent work, Gabbard et al. [45] describe that this mainly sequential process is not applicable to the development of AR interfaces, due to the problem of missing design guidelines or interaction metaphors. Their modified approach emphasizes iterative design activities inbetween the user task analysis phase, where requirements are gathered and user tasks understood, and the formative evaluation phase, where an application-level user interface prototype has been developed and is under examination. This step basically consists of a loop of three components: 1) User Interface Design, 2) Expert Evaluation, and 3) User-based Studies. This loop is intended to influence design and produce lessons learned to be communicated to the community.

A completely different usability engineering strategy is the *Testbed Evaluation Approach* [18], which basically aims to evaluate interaction techniques outside the context of applications under rigorous controlled and isolated conditions in a testbed. The conditions of the test bed are derived from an initial evaluation which produces a taxonomy of the (sub-)tasks, an estimation of the external factors and a selection of performance metrics. It produces a set of results or models which characterize the usability of an interaction technique for the advanced specified task.

The sequential evaluation process by Gabbard et al. [45] focuses mainly on fast application development and makes use of usability evaluation techniques, whereas the testbed evaluation approach by Bowman et al. [18] focuses on low level evaluation and thus makes use of exhaustive human factors experiments (compare Sec. 3.2.2). While both aim for different goals, both have their justification.

3.4.2. A refined Usability/Utility Processes

As a result of the discussion thus far we will now present our concluded usability and utility engineering approach (see Fig. 3.1). This approach brings the fundamental aspects of AR design and evaluation together and refines the existing usability processes. We have just discussed the recent approach by Gabbard et al. [45], which already addresses the issue of the non-existence of AR design guidelines, by putting a loop in the traditional [44] sequential usability approach. Our method refines these existing approaches by looking even more for practical applicability. This is in particular achieved by addressing the fact that highly experienced experts for both the design/execution of experiments and AR rarely exist or, if they do, are mostly not available.

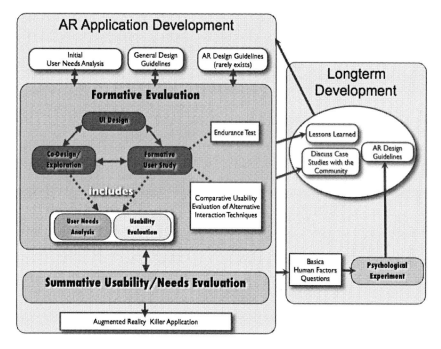

Figure 3.1.: A new developed usability and utility engineering strategy for the efficient development of Augmented Reality applications.

The overall process (see Fig. 3.1) separates the *AR Application Development* from the *Longterm Development*. The *AR Application Development* addresses the iterative development of a specific application and makes use of cost-effective usability/utility techniques *Formative Evaluation* as well as *Summative Evaluation*. The *Longterm Development* focuses on discussing lessons learned and case studies to generate AR design guidelines. It makes use of the rigorous evaluation of human factors (in e.g. Testbed setups [18]) by using techniques of *Psychologically Oriented Experiments*.

The *AR Application Development* is subdivided into the three stages of the classical design process. There is an initial phase which consists of an *Initial User Needs Analysis*, the application of *General Design Guidelines* and, if available, the application of *AR Design Guidelines* to the actual problem domain. The other phases are the classical *Formative Evaluation* and the *Summative Evaluation*. The *Formative Evaluation* is the essential phase of our process, whereas the *Summative Evaluation* is rarely applied. The main goal of this *Formative Evaluation* is to have a cost-efficient method which provides qualitative information for redesign while at the same time educating people about the specifics of Augmented Reality and experimental design. To this end, the *Formative Evaluation* is based on a loop of three elements: 1) the *UI Design*, 2) the *Cooperative-Design/Exploration* procedure, and 3) the *Formative User Study*. With *Cooperative-Design/Exploration* we define the phase in which we only take a few users to explore and reduce the design space in a strong cooperation between test users and the evaluator. It follows a rather ethnographic approach. This helps a great deal to find the largest usability pitfalls and to elicit important user needs. The success of this exploration depends a lot on the test persons. It is important to have test persons from different areas, as discussed in Sec. 3.3.1. This kind of experiment is not to be confused with an informal demonstration. About half of the experiment should follow some formal rules: the evaluator should not lead the participant to the solution, a within-subject design should be used and the participant should actually execute realistic tasks. The other half of the experiment should be a cooperative exploration of different design possibilities. There are two reasons for switching from the *Cooperative-Design/Exploration* to the *Formative User Study*. Either a) we have to solve a problem, or b) we think we have designed a good solution. In the problem case a), we need to find out, for example, which of the interaction techniques is better suited. We then set up an experiment with a few more users to establish this, potentially making use of statistics. In case b), in which we think we have found a good solution, we should verify this in an endurance test. Quite often we will be surprised by the result. Both phases, the *Cooperative-Design/Exploration* and the *Formative User Study*, aim at improving utility as well as usability.

Sometimes situations occur, mostly due to external reasons, where *Summative Evaluations* have to be executed with unfinished AR systems to compare them with an established technology. This is for example the case if the stakeholder wants to know about the specific benefits of a particular technology. Often such situations cannot be avoided since they are explicitly requested by a stakeholder. However, bad performance is to be expected due to the current immaturity of the system. Thus, such evaluations must be conducted with utmost care to be able to give reasons for the likely bad performance of the AR system. We give an example of carefully executing such an experiment in Chap. 5.

3.5. Acceptance of New Technologies

We presented the order picking system (developed in Chap. 5) several times to people from industry (in our lab, at open house days, at conventions and fairs) and thus we were frequently confronted with the question of the acceptance of new technologies. We realized several things. The "nicer" and the more advanced (without being mature) looking the hardware of the prototype was, the more optimistic was the feedback. For example, we started presenting the system on base of the old and toylike Sony Glasstron HMD using self-made paper based shelves, and progressed to using the industrial-style Nomad HMD with real shelves. Correlated with this were users' answers to the question whether the people could imagine to introducing such an Augmented Reality system in their company, or even thinking about working with such a system eight hours a day. Furthermore we collected some anecdotes about managers who tried to introduce mobile computer based order picking systems which were destroyed by the workers, who refused to use them. Other people told us they were thinking about quitting their job, because they were expected to soon work with a computer. Acceptance can be a confounding factor in user studies. Users who do not see a benefit in using a new technology (like Augmented Reality), may for example complain about physical strain [147].

Overall, this means that we have to be careful. It requires some sensitivity when introducing new technologies. It can be a key to increase user acceptance of the final system to let users participate in the design of the systems, e.g., by using them as test persons [78, 27]. An important role is the foreman/manager. According to Heider's Balance theory [57], there exists a balanced triangular relationship between the company (represented by the foreman/manager), the worker and the technology. If the worker has a positive attitude towards the company, then a positive attitude of the company towards the new technology increases the positive attitude of the worker towards the technology [51].

3.6. Conclusion

This chapter introduced usability and utility as highly important goals for the Augmented Reality user interface design. We propose that a user-centered design process is the key for achieving such design goals. However, this requires Augmented Reality user interface experts, design guidelines and evaluation methods. All of this is rarely available for Augmented Reality. Furthermore the Augmented Reality design space is highly unexplored and solutions from one scenario or system are not easily transferable to another one. This guarantees that early designs are filled with usability pitfalls. Additionally, it already requires a lot of skills and experiences to perform user-based studies for classical user interfaces.

These problems are even more pronounced when performing studies with Augmented Reality user interfaces. The often immature and bulky hardware, non-perfect calibrations, inexperienced users and fascination and learning effects can extremely confound the results of an experiment.

As all these aspects come together when trying to evaluate Augmented Reality user interfaces and we rarely have experts for all of them available in real life, we propose a refinement of the classical usability and utility engineering processes. It basically consists of a loop of three elements: design, exploration, user study. The main benefit of this approach is to have a cost-efficient but formal method which provides qualitative information for redesign and at the same time, a possibility for educating people about Augmented Reality and experimental design.

System

Details on the Pick-by-Vision software. Reasons for using specific displays, input devices and tracking systems.

This chapter gives an overview of the Pick-by-Vision software, which was developed and implemented within the scope of this work. Furthermore the hardware aspects, including the tracking system, input devices and the HMDs, will be discussed.

4.1. Software

The software consists of the Pick-by-Vision system itself, a pseudo warehouse management system, which is actually part of the Pick-by-Vision system and a calibration wizard.

4.1.1. Pick-by-Vision System

The Pick-by-Vision system is an application to support the order picking task with HMDs. It is designed for Pick-by-Vison (2D) and Pick-by-Vision (AR) (compare Sec. 2.3). However, its main intention is to guide workers through the warehouse using Augmented Reality visualizations. This functionality is fully described in Chap. 6. Several other features, addressing the important needs of an order picking worker, were implemented on the basis of the domain analysis (Chap. 2) and the evaluations with domain experts (Chap. 5). These features include user management, order handling, error handling (i.e. annotating the different type of errors) and worker guidance before and after the actual picking (for example, to the delivery station). A detailed discussion on the features can be found in Reif [114].

Our first Pick-by-Vision system was an elaborately engineered system. It based on several distributed components, made use of the DWARF middleware [10] and was fully implemented in Java/Java3D. We made use of a highly distributed approach as we initially explored the use of ubiquitous displays [123]. This initial exploration required several computers, in particular to render the different visualizations. As we had to deploy the system frequently in different environments, we replaced the DWARF middleware by a simple but sufficient Java-based network communication. However, we realized the we mostly used the HMD-based visualizations and thus used it in a mobile standalone configuration. In contrast, the tracking system (compare Sec. 4.3) continuous to be distributed. It runs on a wirelessly connected computer. The implementation of the dedicated components follows a Model-View-Controller pattern. The important components will be set out now.

AR User Interface   The AR user interface component is the core of the Pick-by-Vision system. It is responsible for rendering all 2D and 3D elements for the HMD. The most relevant elements and its parameters are adjustable via a configuration file (see Sec. A.1). This configuration file was a key element in the user interface exploration and evaluation, because it allowed us to easily and rapidly change the user interface without making the "expensive" test participants wait. For example, this configuration allows to switch between Pick-by-Vision(2D) and Pick-by-Vision(AR), to switch between the different meta and picking visualizations or to adjust its sizes, offsets, etc. The component itself was designed in such a way that new visualization elements (meta and picking) can easily be integrated.

Warehouse Management   The Pick-by-Vision system needs a spatial description of the warehouse (see Sec. 4.1.2) and a management of the stock and the orders (see Sec. 4.1.3).

Tracking   The Pick-by-Vision system has a direct connection to the ART Dtrack system (see Sec. 4.3). Furthermore it has an Ubitrack [69] component, which makes it possible to use a wide range of tracking sensors or even to make use of sensor fusion.

User Input   The different user input modalities, which are handeld via the user interface component, are described in Sec. 4.2.1.

Logging   Every user input and state change is logged with a timestamp to enable detailed analyzes of the user performance.

4.1.2. Warehouse Layout

The warehouse configuration, in particular the layout of the shelves, is described by a XML-file (see Appendix A.2). It follows the structure of a real warehouse (see Fig. 5.25), which consists of aisles. An aisle has at least one shelf, but most have two opposing shelves and can further have further shelves above each other or in a row. Each shelf has a local coordinate system, as it is only the shelf which is moved within the tracking setup whereas the inner parts of a shelf are usually rigid. Thus, if several shelves of the same type are used, only the origin of a shelf's local coordinate system has to be adjusted. We typically choose a point on the shelf as the origin, which could be best estimated or most practically within the tracking setup. A shelf itself consists of a matrix of boxes/storage locations. This matrix is simply described by horizontal/vertical distances and the size. Each shelf has a size, which specifies the location and dimension of the "invisible" occlusion objects.

For scenarios which require a high degree of accuracy between the virtual and real warehouse layout, we developed an option to render the occlusion model "visible". This enables the developer to walk through the warehouse (wearing the HMD) and to compare the real warehouse with its virtual model. Fig. 4.1 shows such a view through the HMD. We use a ruler to measure the deviation between real warehouse and virtual model. This deviation can thus easily be corrected by adjusting the configuration file or by slightly moving the shelves. This procedure requires some AR-experience, but allows a reduction of the overall error of the augmentation's position.

Figure 4.1.: Fine adjustment of the shelves, photographed through the HMD. It shows a ruler measure and minimize the deviation between virtual and real shelf.

4.1.3. Warehouse Management System

Warehouse Management Systems (WMS) are used to control the movement and storage of goods within a warehouse. Even though the integration of a "real" WMS was intended, it did not take place due to lack of resources at the cooperating WMS provider. Thus the Pick-by-Vision system is based on a pseudo WMS consisting of two configuration files which was completely sufficient for this work. The configurations are the list of items (see Appendix A.3) and the list of orders (see Appendix A.4). This WMS associates the list of items with the warehouse layout configuration (described in Sec. 4.1.2) to calculate the spatial information for each storage location. The items can have additional features, which can be optionally used for information presentation in the HMD (e.g. an image of the item or the textual description of the location). The actual orders consist of a list of IDs (linked to the item list) and the corresponding amount.

4.1.4. HMD Calibration Wizard

Two steps have to be performed for the HMD-calibration: The optical see through calibration and the adjustment of the "Nose Pose".

For the optical see through calibration of the HMD we implemented a wizard, which is based on the Single Point Active Alignement Method (SPAAM) [144]. The method requires the user to sequentially generate about 20 2D/3D correspondences, which means the user has to align 2D points shown in the HMD with a measurement tool which is placed somewhere in the 3D space. This approach calibrates extrinsic and intrinsic parameters at the same time. The extrinsic parameters basically describe the spatial relationships between the optics of the HMD and the attached marker target (compare 4.3) . The intrinsic parameters describe the optical characteristics of the display itself. In particular, those characteristics are dependent on the position of the user's eye in front of the display. Theoretically this means that the display has to be re-calibrated or at least re-adjusted [48] when it only slightly moves on the user's head, or in particular, when another user wears it. Practically, we never re-adjusted it when the users changed, due to the following practical reasons. First of all, the HMD we mostly used, only allows users to see the virtual image from a certain perspective. Secondly we tested more then 100 users in formal experiments (and probably even more informally), which ment that all users would have had to execute the re-adjustment procedure. Even worse, we would have needed to train them how to carry out a good calibration. A drawback of the SPAAM method is that it cannot compensate for non-linear distortions. This leads to inaccuracies in the corner regions of the optical see-through display, which sometimes produces pillow-like images, rather than rectangular ones. Nevertheless, the accuracy of the whole system was sufficient for our experiments and was rather influenced by the wide area tracking setup than by a badly calibrated HMD. Setups which need a better calibration method, should implement the Display Relative Calibration [107]. It makes use of a pre-calibration procedure and thus requires only a minimal input from the enduser and additionally calibrates the radial distortion effects.

The second calibration which has to be performed is the estimation of the nose pose. This is a position about 20 cm in front of the user's nose. The nose pose is the origin of the user-centered coordinate space where in particular all meta visualisations (compare Chap. 2.3.2) are placed. The nose pose is estimated by recording the 6DOF position of a tracked measurement tool, which is placed in front of the user's nose, while the user wears the tracked HMD.

4.2. Hardware

The mobile Pick-by-Vision hardware basically consists of the input and output device (i.e. the HMD) and a mobile computer running the software. Additionally it makes use of an external tracking system (see Chap. 4.3) which was connected via WiFi to the mobile tablet PC. The typical setup can be seen in Fig. 4.2. The mobile PC, which was used in most of our setups, was a 13-inch, 2kg tablet PC (Core 2 Duo 2GHz and a standard 3D Graphics card). The users carried it in a small backpack. Even though

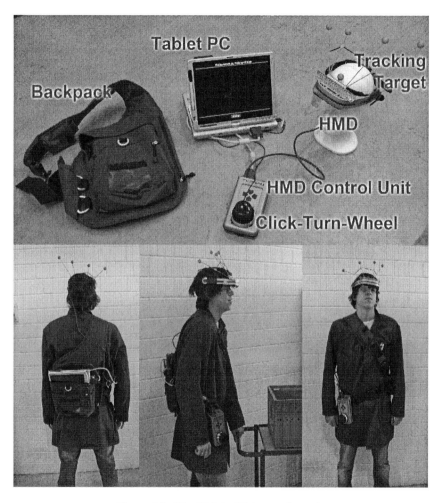

Figure 4.2.: The Pick-by-Vision equipment.

the number of components seems quite large, the subjects did not complain about the system's weight or dimensions (even when wearing it for two hours). In the following sections the different input and output devices will be discussed.

4.2.1. Input Devices

There are only a few user inputs required by the Pick-by-Vision system. At the first stage of development those were: requesting an order and getting the next order line. During the later development this was extended by functions to log in, getting the previous order line, annotating an error (including its kind) and terminating an order.

At the first development stage we did not have an ideal input device and so we used the WizardOfOz [17] method, which is a common technique in the evaluation of 3D user interfaces, in particular to avoid confounding the experiment because of the test participants' use of an immature input technique. Practically, this means the user says, for example, "I have picked an item" and the assistant presses a mouse button. Therefore we had client software running on a stationary computer, being WiFi connected with the actual Pick-by-Vision system in the backpack of the user.

In the later development stage we integrated a click-turn-wheel knob which is the only input device to the system, allowing for four inputs: turn (left/right), short click and long click. Such a device was recommended by [39, 11]. It was mounted on top of another control unit, which is attached to the belt. In the most evaluations of this work we used the "PowerMate" device from Griffin Technology. However as this device had reliability problems we later replaced it with a similar device, the "Space Navigator for Notebooks" from 3D Connexion.

A future system should make use of voice entry for the system control. Reif et. al. [115] have shown the benefits of using voice entry for Pick-by-Vision systems: by forcing the user to verbally acknowledge the number of picked items, the errors made with the system were reduce (in comparison to a Pick-by-Paper system).

4.2.2. Output Device - HMD

Besides some debugging windows on a normal computer screen, the actual output device for the user is a head-mounted display (HMD). There are several other ways to provide a mobile information presentation in warehouses and we even explored those [123]. However the scope of this thesis is restricted to HMD based solutions.

Basically there exist two types of HMDs: video see-through displays and optical see-through HMDs. Video see-through displays use small monitors in front of the user's eye(s). Those monitors are used to show an augmented video of a camera, which is mounted in front of the video see-through display. Thus video see-through HMDs in most cases completely seal the user from reality, that is because they show the user only an augmented video which is always a little bit behind the reality and at the same time reduces the natural field of view, as there is no lightweight full field of view display available. Optical see-through HMDs make use of an optical combiner to realize the Augmented Reality view: E.g. a half mirrored glass plate allows the view on display and reality at the same time. This implicitly ensures that the user can always see his or her reality in real time, which is quite an important factor. Taking into account that the user has to walk and navigate safely in an industrial setup, we cannot use a video see-through HMD. In several discussions industrial experts agreed with this.

However optical see-through HMDs have an accommodation discrepancy problem, which will be discussed below (see Sec. 4.2.2). Additionally, it is not easy to choose the right HMD; even though from time to time some new concepts are proposed, there is no perfect HMD available at the market: Mostly they are too heavy and bulky, the field of view is too small, the display is not bright enough, etc. There are several discussions about the right HMD [104, 114] to be used in industrial setups. At this point we will briefly explain why we used the following two HMDs. In the first experiments we used a binocular Sony Glasstron HMD (compare Fig. 5.3), simply because during that time it was the best one available in our lab.

During the first test we realized, that the Sony Glasstron HMD heavily occludes the peripheral field of view and actually only allows a tunnel vision like view with a quite dark display. For this reason we decided to choose another display, which had to meet the following requirements: bright, minimal occlusion of the peripheral field of view, lightweight and a maximum of field of view. We did not explicitly look for a binocular or stereoscopic display, as our intention was to permit the user as much of his natural field of view as possible. Furthermore as much as 20% of the population is actually stereo-blind [151]. We decided to use the Microvision Nomad ND2000 VGA HMD, which makes use of a direct retinal projection to generate the image. It therefore only requires a dimple glass plate as optical combiner. This dimple glass plate does not have a frame, as contrasted with most other displays (particularly the Sony Glasstron) and it therefore does not cut the user's field of view. Furthermore it uses a red laserprojector which produces a relatively bright image, compared to other displays. Consequently this HMD can only produce monochromatic images. Its actual field of view is $23°$x$17, 25°$ ($28, 75°$ diagonal), which is relatively small but compares with other HMDs available in the market. It has a resolution of 800x600 ($0, 036°$/Pixel) at 60Hz, weighs 128g and, importantly, its battery runs 8 hours. Another advantage is that it does almost look like a fancy high-tech device and thus the acceptance rate of our test users seemed to be higher than it was for the Sony Glasstron. However, this device does not seem to be perfectly suited as a few participants complained about sore eyes after using it for a prolonged period (compare Sec. 5.7.3). We are not sure whether this was due to the display type (retinal projection) or due to other reasons (such as the monocular usage), or because of having to switch focus between real world and focus distance of the display frequently. This problem will be discussed in the following section.

Adjusting the Focus of the Nomad HMD

The problem of the accommodation discrepancy for the use of optical see-through displays (like the Nomad HMD) for Augmented Reality is widely known [85, 71, 70]. The user of an HMD can only focus on the virtual image of the display and the real world simultaneously when both are at the same distance [30], which is rarely the case. Rolland et. al. [119] stated that it can be shown that rendered depth errors are minimized when the virtual image plane is located in the average plane of the 3-D virtual object visualized. As a solution to the various conflicts in accommodation, Rolland et. al. [120] suggest to allow autofocus of the virtual image plane as a function of the location of the user's gaze point in the virtual environment, or to implement multifocal planes. A frequent change of accommodation produces fatigue, but placing the virtual focus in typical handling distances of 0.6 meters forces a continuous contraction of the ciliary muscle [104].

Since currently none of the proposed autofocus/multifocal plane HMDs are available, we have to deal with the single feature of our Nomad HMD allowing to manually adjust the virtual image plane between 0.3 meters to infinity. The users of our order picking system have to regard two types of visualizations: 1) the 3D augmentation and 2) 2D textual information. The augmentation consists of the meta navigation in combination with the picking visualisation in front of the box and can stretch over a focus area of about 0.4-3.5 meters. On the other hand the user needs to read text from the HMD in different situations, but the most relevant situation seems to be when the user holds the picked item, which is typically at a distance of about 0.6m. At that point the user has to compare the article number shown on the HMD with the article number printed on the article itself. This requires a multiple change of focus (between HMD and reality) in a short time period, as the number is usually quite long. Finally, we put the focal distance of the HMD in the middle of the discussed values, somewhere between 1.0 and 1.5 meters, as the Nomad does not allow an exact adjustment.

4.3. Tracking System

The tracking system is required to estimate the position of the user (in particular the position of the HMD) in the warehouse. For that reason a "large area" tracking system is required which can estimate the full 6 degrees of freedom (6DOF - 3 translation and 3 rotation). This tracking system has to be at a high level of maturity (continuous/robust tracking, light-weight marker targets, low latency, high update-rate and an accuracy in the area of a few millimeters) so that it is possible to execute experiments without interruptions and to not confound the results of the experiment. The ART DTrack system, which was available in our labs, fulfilled these requirements. The only disadvantage is, that it needs relatively bulky marker targets in larger setups to ensure accuracy. This had to be attached to the HMD as can be seen in Fig. 4.2. However we adjusted the design of these markers, in such a way that the users did not bump the marker into the shelves. Altogether, this is not a solution which is suitable for the continuous operation under industrial conditions, but it perfectly allowed us to develop the Augmented Reality user interface.

4.4. Warehouses

The Pick-by-Vision system was developed and used in three different warehouses. The first exploratory experiment was executed in an area of $1.5m$ x $3m$ with boxes only in the height of the upper body (see Fig. 5.3). Most of the evaluations took part in the small one-wall warehouse consisting of 96 small boxes (see Fig. 5.7). Later we went to a "real" warehouse consisting of two aisles, each 3.4 meters long and 1 meter wide (see Fig. 5.25). The shelves consisted of two layers from 0.4 meters to 2 meter with 280 stock locations (boxes).

Evaluations

The iterative process of (re-)designing and evaluating the Augmented Reality visualizations and the Pick-by-Vision system itself

Many of these theories have been killed off only when some decisive experiment exposed their incorrectness... Thus the yeoman work in any science [...] is done by the experimentalist, who must keep the theoreticians honest.

MICHIO KAKU

This chapter presents the formative process of developing the Pick-by-Vision system and thus is the main contribution of this thesis. It is separated into 8 steps describing the consecutive user evaluations and improvements of the system. This process started in the lab and ended in an almost realistic setup in a factory hall. Fig. 5.1 gives an overview of this process and thus of this chapter. In Sec. 5.1 we started with a huge setup to explore the usage of HMD-based AR and other display technologies to support the picking process. Sec. 5.2 describes our search for a more convenient AR visualization – as in the first experiment some users picked out of the wrong storage locations. The result of this search is then tested in Sec. 5.3. As the Box Finding seemed to be working after these experiments, we concentrated on finding a better Meta Visualization to guide the user when the Box Finding visualization is not in its field of view (Sec. 5.4). The resulting visualization was then brought from the lab to a warehouse (Sec. 5.5). In Sec. 5.6 we describe the first test against other technologies in the warehouse. This resulted in an adaption of the meta visualization, which we then underwent in a two hour stress test (Sec. 5.7). After this test we made some fine adjustments of the meta visualization, which were proved in a last experiment, described in Sec. 5.8.

Each section follows a basic structure to describe the current iterative step of the evaluation. According to that, each section basically consists of the following three categories, which are summarized in a table at the beginning of each section:

- *Objectives* – What was the reason and the purpose for executing the experiment?

- *Experimental Setup* – How was the Experiment executed and which experimental methods were used?

- *Results* – What are the results of the experiment (for the visualization, the system and what are the lessons learned related to the evaluation of AR Interfaces)?

The results for visualization and system had an impact on the improvement of the Pick-by-Vision Systems itself, whereas the Evaluating AR/ Lessons Learned category had a huge impact on Chap. 3 and the experimental setup of the following evaluations. As Chap. 3 is basically a result of this chapter, this chapter cannot perfectly follow the advice given in Chap. 3.

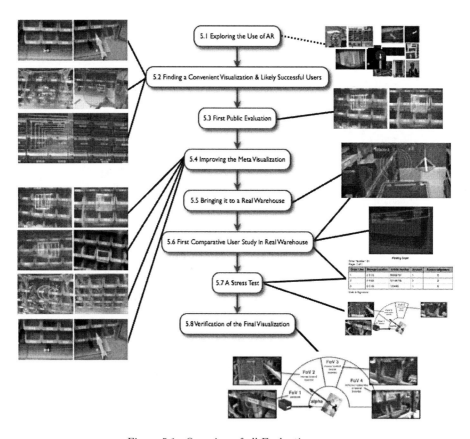

Figure 5.1.: Overview of all Evaluations.

5.1. Exploring the Use of AR to Support the Order Picking Process

Objectives		Exploring different visualizations on different displays to support: a) Aisle Finding (Coarse Navigation) b) Box Finding (Fine Navigation)
Experimental Setup	Hypothesis	Informal expectation: Users perform faster and make less errors when they were provided with 3D information than with 2D and 1D Information
	Independent Variables	Type of Visualization (1D, 2D, 3D/AR), Type of Display (HMD, stationary Monitors, PDAs)
	Dependent Variables	Major Variables: Time, Failure
	Test Method	Explorative experiment, within-subject design, warehouse 1
Results	AR - Visualization	Subjects made more errors and did not perform faster when using AR-based visualizations.
	System	--
	Evaluating AR / Lessons Learned	Novices need a huge amount of time to become comfortable with AR and thus have a higher influence on the test than the different levels of the variables.

Figure 5.2.: Summary of the first explorative experiment to test different visualisations for the order picking process.

Our first approach toward supporting the order picking process with computer visualizations was to compare the different possibilities of presenting working instructions to mobile workers in an explorative experiment. To this end, we classified three types of displays and three types of visualizations, resulting in 3x3 = 9 presentation schemes. We took an instance of each display type (HMD, stationary monitors, PDA) and implemented for each type of visualization (1D, 2D, 3D) an instance on each individual display. The second factor influencing the setup of this experiment originates from the picking process itself. As already described in the domain analysis (see Chap. 2.3), we have to support the two essential picking tasks *Aisle Finding* and *Box Finding*. This basically resulted in the fact, that we had to set up two individual experiments, one for the Aisle Finding and one for the Box Finding – resulting in 2x9 = 18 different treatments.

However, after the completion of this experiment, we realized, that the main failure was to test ad-hoc built system setups. Which means, that all visualizations were tested in a rather premature environment. This resulted in the fact that we found a lot of usability problems for each individual technology, but did not gain valuable insights by way of a comparison between the different technologies. Using this setup we showed which visualization was at a better stage of development rather than showing which type of visualization is better. In retrospect, this was a beginner's mistake (which is probably and sadly made quite often in scientific evaluations). Due to this, we only present the part of the experiments relating to the HMD-based visualizations, because the rest of the experiment did not have an impact on the following evaluations. Fig. 5.2 gives an overview of the experiment. This work and the results for the other displays were first presented in [123, 40].

5.1.1. Objectives

The motivation of setting up this experiment, was to get insights into the use of different computer visualisations on different displays to support the order picking process in general. However, we were driven by the idea of showing that 3D-based visualizations and in particular an Augmented Reality-based visualization, can outperform all other ways of presenting information. That was the reason for setting up a comprehensive system consisting of different kinds of displays and different kinds of visualizations.

5.1.2. Experimental Setup

The first version of our experimental warehouse setup consisted of selfmade carton-based shelves(see Fig. 5.3 b) . Each box – storage location – in a shelf had a size of 25cm×15cm (width×height). For the Aisle Finding experiments, the warehouse consisted of 4 double-sided shelves resulting in 5 aisles to go. The warehouse for the Box Finding experiment consisted of a single shelf, which was 5 boxes wide and 10 boxes high. The shelves were numbered sequentially.

The mobile hardware setup is described in Chap. 4.2. For the visual output we used the Sony glasstron HMD (see Fig. 5.3 a and Sec. 4.2.2). As our first intention was to gain an insight into the use of AR for mobile workers, rather than developing a completely mobile picking system, we used the quite popular WizardOfOz method for the user input (see Sec. 4.2.1). This means users basically feign the user input (e.g. "I picked it"), rather than pressing a real button.

Independent Variable for each experiment: Type of Visualization at 3 levels

For both experiments (Aisle Finding and Box Finding) we set up three different visualizations clustered by their visual dimension (1D, 2D, 3D). The 1D visualization for both experiments was a text with the shelf/box number. For the 2D visualization we displayed an abstract line-based (you-are-here) map highlighting the target box/shelf. The map was oriented horizontally for the Aisle Finding and vertically for Box Finding.

Figure 5.3.: Setup of the first evaluation a) Sony Glasstron HMD, with retro reflective markers (to be tracked by the ART DTrack System). b) Four double sided shelves for the pathfinding test.

However, in retrospect the most interesting were the 3D (Augmented Reality) visualizations. As discussed in Sec. 2.3 we need individual visualizations for the Aisle and Box Finding. Furthermore both visualizations have to be supported by a Meta Navigation visualization. The three visualizations will now be discussed next:

Meta Navigation   We used a *Compass-like Arrow* visualization to tell users where to turn their heads when the picking target was outside their field of view. The Compass-like Arrow consists of an arrow at a fixed distance of about 20 cm in front of the user's nose. The arrow points to the next relevant augmentation and is extended with a rubber band of flexible length. Three fins are attached to the back end of the arrow, which provides a better depth perception than just using an arrow consisting of a cone and a cylinder [141] (see Fig. 5.4b and 5.5b).

3D - Aisle Finding Experiment   For indicating the target shelf, we set up a virtual traffic sign in front of the shelf with the next target (see Fig. 5.4a). The visualization exploits an occlusion relationship between the shelves and the sign as an additional depth cue. The traffic sign could be seen from everywhere, as all shelves had a maximum height of 140 cm.

3D - Box Finding Experiment   We implemented a 3D Arrow that slightly immerses into the box of the next target (Fig. 5.5a). This reinforces the effect of depth perception

 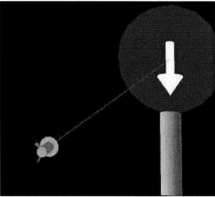

Figure 5.4.: Aisle Finding visualization a) A photograph through the Sony Glasstron -
The shelf partially occludes the pillar of the traffic sign b) The rendered
image

provoked by the occlusion of the shelf, when seen from an oblique angle. The arrow itself
is similar to the one used for the Meta Visualization.

Hypotheses

We expected users to be faster when they were provided with AR information than with
2D and 1D information. However, since this was only an explorative experiment, we did
not set up a proper hypothesis [15].

Test Method & Dependent Variables

We designed two separate experiments for the *Aisle Finding* and *Box Finding* tasks. The
setup was within-subject and the participants performed the tasks in permuted order.
We measured (dependent variables) *Way Time + Dead Time* for the Aisle Finding
evaluations and *Picking Time + Dead Time* for Box Finding evaluations. Additionally,
we investigated the number of picking errors the participants made. The test participants
were observed during the tests and had to fill out questionnaires afterwards.

Aisle Finding Test   The participants had to go to a specified column of the shelve and
say "I have arrived" (without picking an item), always starting from an initial position
in the lab. Each basic batch of tests consisted of 3 pathfinding orders listing 6 positions

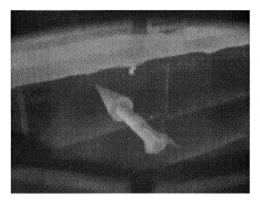

Figure 5.5.: Box Finding visualization of the first test: a) Large arrow in front of picking target. b) Meta Visualization: Compass-like Arrow in front of the user's nose extended by a rubber band.

(i.e. 18 operations) for each visualization. For each display the same batches were used, permuted for different persons.

Box Finding Test   The test participants had to pick a specified number of items from the correct box on a shelf and then say "I picked". Each batch of tests consisted of 3 picking orders listing 10 items. Test participants had to run a batch on the HMD for each visualization. Test participants thus were requested to make $3 \times 10 = 30$ picks.

5.1.3. Results

Eighteen test participants (age 23-48) participated in the experiment. For Aisle Finding, we could not show any differences in time between tests with 1D or 2D visualizations, while the 3D visualization was several seconds slower than the 1D and 2D visualizations. We observed two interesting effects: Firstly, the test participants were almost 50% faster in the third cycle than in the first. This was probably caused by learning effects and by a fascination of novel AR users. Secondly, there were two distinct groups of fast and slow users. The sub-clusters correlated with users who had already prior exposure to computer games or AR/VR applications and those who did not. There were only a few Aisle Finding errors, mostly when test persons had misunderstood the initial instructions.

In the Box Finding test, we could not measure significant speed differences between the 1D, 2D and 3D visualizations. However, failure rates showed clear differences. Users made up to 10 times more mistakes with the AR-based 3D visualizations in the HMD than with the 1D/2D visualization. In most cases, users picked items one row too high or too low, indicating a problem with depth perception using AR. We showed AR-based picking later at a small fair using larger shelves (boxes $40cm \times 45cm$ ($width \times$ height) rather than $25cm \times 15cm$). Here, people made almost no mistakes. That means the AR-based system had a lower bound for the box size. Below this limit, box identification became ambiguous. Comments in the subjective questionnaires suggested this was partially due to the visualization scheme and partially due to the insufficient optical see-through quality of the Sony Glasstron HMD.

AR - Visualization

The two main observations with respect to the AR-based scheme were: A) People often picked items from the wrong box (one row too high or too low), indicating a problem with perceiving the depth of the displayed 3D Arrow (see Fig. 5.5a), and B) people needed some time to familiarize themselves with Augmented Reality altogether. The test participants were almost 50% faster in the third test cycle than in the first, due to learning effects and the high initial general fascination with Augmented Reality of first-time users. After having overcome the obstacle of becoming comfortable with Augmented Reality, the test users mentioned that they found the metaphors intuitive. People who were unexperienced with new technologies such as augmented/virtual reality, needed even more time to use the AR-based system efficiently.

Evaluating AR / Lessons Learned

We learned, that one should not simply setup a huge experiment with several variables, a lot of ad-hoc implementations and a huge amount of users. We had a lot of trouble with usability problems in the ad-hoc implementations. We learned the lesson that one needs a lot of practical experience in setting up, executing and analyzing such an experiment, which we did not have at this stage of the work. All these facts influenced Chap. 3.

5.2. Finding a Convenient Visualization and Likely Successful Users Of AR Systems

We discussed the results of the first test series with several logistics experts and workers. They encouraged us to focus on AR based visualizations rather than the other ways of information presentation. They further told us to primarily investigate methods towards preventing users from making errors, i.e., striving for a *zero error* system, and thus concentrating on the Box Finding task.

A problem, arising from the last experiment (Sec. 5.1), is the long learning curve of novices and our supposition, that different personalities could influence this learning curve and respectively the general performance with AR-Interfaces. To this end, we exposed different psychological tests to the users. Fig. 5.6 gives an overview of the whole experiment. This experiment was originally published at [124, 125].

Objectives		a) Focussing on AR support for Box Finding (evaluating alternatives to the Compass-like Arrow ->(Frame/Tunnel)). b) Are there likely successful users of AR systems?
Experimental Setup	Hypothesis	a) Hypothesis: there will be differences between the visualizations according to time and failures. b) Exploratory question: can we predict the performance of a user with the Pick-by-Vision (AR) system?
	Independent Variables	Type of visualization: Arrow (supported by Compass-like Arrow), Frame (supported by Compass-like Arrow), Tunnel
	Dependent Variables	Time, failure, IQ, personality type test (NEO PIR), performance in a computer game
	Test Method	Formal experiment, within-subject design, warehouse 2: 96 small boxes on one wall, Nomad HMD, psychological questionnaires
Results	AR - Visualization	Users were faster and made no errors, when using the Frame visualization. The Tunnel seems to be a good Meta Visualization, but not a good Box Finding visualization.
	System	The Nomad HMD seems to be better suited for AR-Picking than the Sony Glasstron HMD.
	Evaluating AR / Lessons Learned	An elaborate try-and-ask phase is required to compensate for learning effects (in particular of novices and different types of personality types).

Figure 5.6.: Overview of the experiment to find a convenient picking visualization and likely successful users of AR systems.

5.2.1. Objectives

With this experiment we try to find a visualization to support an error free and fast Box Finding. Furthermore we looked for different criteria (e.g. personality types, IQ, performance with computer games), which identify likely efficient users of Augmented Reality interfaces.

5.2.2. Experimental Setup

For setting up this experiment, we replaced the previously used Sony Glasstron HMD with a Nomad HMD (Fig 5.7a) to benefit from the very high see-through capabilities and less occlusion of the peripheral field of view of the Nomad. Further reasons for the choice of the Nomad HMD are given in Sec. 4.2.2.

We also refined the experimental warehouse setup. Users now were placed directly in front of a shelf with small boxes (Fig 5.7b). The new shelf is 12 boxes wide and 8 high (= 96 boxes). The boxes have a size of about $10cm \times 10cm$ and are thus smaller than the boxes in the first experiment ((Sec. 5.1)) .

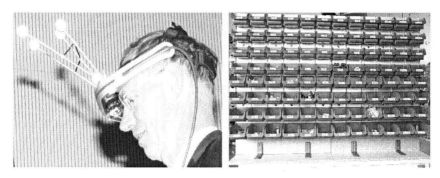

Figure 5.7.: The experimental setup of the second generation. a) Nomad from Microvision uses a small glass plate in the field of view b) The new shelves which are much smaller than the shelves in the first test.

Independent Variable: Type of Box Finding Visualization at 3 Levels

After executing a literature review and completing discussions with several users and experts, we decided to implement the following three visualizations to support the Box Finding. Each visualization has to address the problem of Meta Visualization to guide the user when the object is not within the field of view.

The Arrow   As a reference visualization, we used the Arrow from the first test (Fig. 5.5), in combination with the Compass-like Meta Visualization. Fig. 5.8 shows how this looks through the Nomad HMD.

Figure 5.8.: The arrow in front of a box (photographed through the Nomad HMD).

The Frame   Several test participants recommended using a "simple" highlighting of the box by a rectangular frame. The virtual Frame is supplemented by the Compass-like Arrow, the same Meta Visualization we use to support the Arrow. The implementation of this frame can be seen in Fig. 5.9.

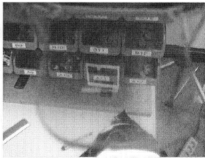

Figure 5.9.: The frame highlighting a box (photographed through the Nomad HMD). The frame is partly occluded by a virtual occlusion object of the shelf. The Frame is supported by the Compass-like Arrow visualization.

The Tunnel A tunnel metaphor is a well known 3D concept for guiding pilots using head-up displays in aeroplanes. Lynda et al. [77] state that it does not matter whether the individual elements of the visualization symbolizing the tunnel are connected or not. Biocca et al. [14] adapted this metaphor to Augmented Reality-based picking tasks. They used a tunnel to visibly link the head-centered coordinate space directly to an object-centered coordinate space. They designed their tunnel by aligning elements on a Bezier curve between the HMD and the object to pick. Their user study showed, that picking by using a tunnel is much faster than just highlighting an object in 3D. Thus, they basically found an alternative to our Compass-like Arrow metaphor for directing the user's attention while using an HMD with a small field of view. However, in their use case they did not need exact navigation as the objects in their experiment were further apart than the objects in our experiments. In both scenarios (flight navigation [77] and picking [14]) the tunnel was found to be a preferred solution for rapidly guiding the user in 4π steradians. Furthermore, it has minimal attention demands and minimizes the mental workload.

To this end, we implemented the tunnel to indicate the box to pick from. Technically we aligned objects on a Bezier curve going from the front of the HMD to the box. Actually, it looks like attaching a flexible hose of a vacuum cleaner from the HMD to the object. As this visualization includes the Meta Navigation, we did not use the Compass-like Arrow in this case. Our implementation is shown in Fig. 5.10.

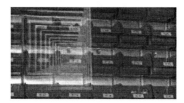

Figure 5.10.: The Tunnel visualization (photographed through the Nomad HMD). The blue shadow originates from the optical combiner of the Nomad.

Choice of Personality and Performance Tests

As described at the beginning of this section, this experiment was not only meant to give information on finding a better visualization. We also looked for different criteria to identify likely efficient users of Augmented Reality user interfaces. Inspired by a work of Howe et. al [68], who tried to identify likely successful users of Virtual Reality systems, we looked for adequate (psychological) measures, which correlate with an efficient use of AR-Systems. Howe et. al [68] basically used personality type indicator tests and performance tests (e.g IQ Tests), which are quite popular for pre-employment testing. Their conclusion was that competence and a strong temperament make a "likely successful" user of VR systems. After discussions with psychologists, we decided to extend the way of testing by measuring the users' performance with conventional 3D computer games. Finally we decided to use the following tests.

Big Five - NEO-PI-R We used the NEO-PI-R [106] test to classify the users' personalities. It is a successor of the Myers-Briggs Type Indicator [94], which was used by Howe et. al [68]. The NEO-PI-R tests for the Big Five (based on the OCEAN model): Openness to Experience/Intellect (close/open), Conscientiousness (disorganized/organized), Extraversion (introverted/extraverted), Agreeableness (disagreeable/agreeable), and Neuroticism (calm/nervous). The NEO-PI-R consists of a questionnaire of 240 items to rank.

I-S-T 2000 R As an IQ test, we choose the I-S-T 2000 Test in the revised version [82]. This test does not simply measure the overall IQ, but delivers detailed results for the figural/numerical/verbal intelligence, the figural/numerical/verbal knowledge and the memory. The average test duration including introduction is about 100 minutes.

Questionnaire A general questionnaire was designed to gain further insights. It asks questions relating to age, gender, profession, hobbies and users' experiences with computers and AR/VR.

3D Gaming Test We designed two 3D computer gaming tests, where participants had to perform tasks in virtual reality environments. In the first setup, the participants had to run through a level of an ego-shooter (compare Fig. 5.11). The path to follow was indicated on a printed map. The second exercise was a 3D maze, where the test participants had to fly in 3 dimensions through wireframe cubes to find a distinct target cube - compare Fig. 5.11. The challenge in both cases was to perform the tasks as fast as possible.

Hypotheses

After having presented our independent variable, the type of box finding visualization at its three levels (Arrow, Frame, Tunnel), we now want to show a difference between the levels. To this end we use the dependent variables of task performance time and errors. As we did not have an idea in advance whether the new visualizations (Tunnel or the Frame) would be better or worse than the Arrow, we just set up two undirected hypotheses, that there will be a difference between the visualizations according to time and failures: $H_{1-Error-Exp2}$ and $H_{1-Time-Exp2}$.

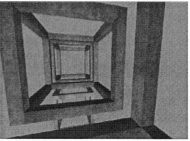

Figure 5.11.: Screenshots from the test applications. a) A simple 3d Shooter (without enemies) b) A 3D maze - the green cube is the way out.

$H_{1-Error-Exp2}$: There will be a difference in the error rates according to the type of visualization.

$H_{1-Time-Exp2}$: There will be a difference in the task completion times according to the type of visualization.

During the analysis of the experiment, we had to reject the respective Null-Hypotheses $H_{0-Error-Exp2}$ and $H_{0-Time-Exp2}$, that there are no differences to be able to accept the alternative hypotheses $H_{1-Error-Exp2}$ and $H_{1-Time-Exp2}$.

As secondary questions within this experiment, not yet formulated as a real hypotheses, we looked for different criteria and tests to make it possible to identify likely efficient users of Augmented Reality interfaces. To make it measurable, we tried to find a correlation between the participants' results of the Big Five personality test, the performance in the structured IQ test, the performance with the 3D gaming tests and the overall performance with the Pick-by-Vision system in general.

Test Method and Dependent Variables

The setup of this experiment mainly consisted of the following two parts: A) the personality and performance tests and B) the actual Augmented Reality Box Finding test.

The personality and performance tests (IST-2000-R (IQ-Test) and 3D Gaming Test) were performed in groups/ individual sessions on different dates, the NEO-PI-R and the general questionnaire were mostly executed as homework. The test participants had to execute all tests as described in Sec. 5.2.2. Hence, the dependent variables of phase A are the IQ and Big Five (described above), and the average task performance time for both 3D gaming tests.

The second part of this experiment was the actual Box Finding test. As described above, the single *independent* variable was the visualization metaphor, providing three levels: Arrow + Compass-like Arrow (Fig. 5.8), Frame + Compass-like Arrow (Fig. 5.9), and Tunnel (Fig. 5.10). For this part each test participant spent about 30 minutes.

For the test, the users were asked to stand in front of the shelf and to pick items. It was not possible for them to see the entire shelf while standing in front of it. Hence people had to move their head to see the boxes to pick from - thus they had to make extensive use of the Meta Visualization. As the test participants did not need to move a lot, we did the test without making the people wear a backpack. Instead we connected the HMD by wire to the server. At startup, each test participant received an introduction with three items for each visualization to play around with and to be able to ask questions about anything they did not fully understand (*try-and-ask phase*). We did this specifically to compensate for the effect of fascination, which gave us a high degree of variability in the measured results in the first experiment (Sec. 5.1). The test participants had to perform 3 orders with 9 items for each visualization. For each of the three visualizations we used the same orders, which was not obvious to the test participants. The order in which the test participants had to use the visualizations was permuted to compensate for learning effects. The test participants had to start each order with their backs to the shelf. Upon a start signal they turned around and executed the test. For system control we used the Wizard of Oz technique, i.e., people had to say "I picked it!" whereupon we switched to the next item location. The system logged the time at this point and a simple harmonious sound from the speakers indicated the change in visualization. We learned from former experiments that the sound feedback is quite important in order not to confuse the test participants, when they do not press the button by themselves. The main dependent variables of this test were task performance time and errors made (wrongly picked items).

5.2.3. Results

We used 34 test participants between 15 and 49 years (mean age: 27.3, StdDev: 5.8, 24 men, 10 women) in a within-subject design. Half of them were from the campus the other half were from the city. Each of them received some candies (the picked items) as reimbursement. In the following part we analyze the measurements of the dependent variables Errors and Picking time and discuss further observations.

Errors

We recorded 33 valid measurement series for the errors. From this, the error rate was calculated (wrong order lines/all order lines – see Sec. 2.1.4). The mean error rates amounted to 0.0550 (StDev: 0.10232) for the Arrow, 0.034 (StDev: 0.01424) for the Tunnel and 0 for the Frame. However, a visual inspection of the structure of the error rates (Fig. 5.12) and the absolute values (Arrow $n = 49$, Tunnel $n = 3$, and Frame $n = 0$) give a clear result and make further statistical analysis superfluous. The users

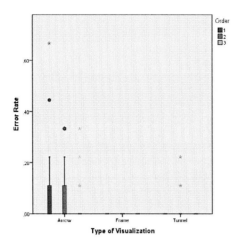

Figure 5.12.: Box plot (median, quartil, min/max, outliers) of the error rates (according to the order) in the second experiment.

had serious problems with the Arrow, at least in the beginning, the Tunnel produced a few problems and the Frame was perfect. Thus we can reject $H_{0-Error-Exp2}$ and accept $H_{1-Error-Exp2}$.

Picking Times

We had 32 valid measurements for the time. We did a pre analysis of the picking times to check for learning effects between the different order sequences the test participants executed with each visualization. This was done, because in the previous experiment (Sec. 5.1) users needed up to 50% more time for the first order than for the last. To this end we drew box plot (Fig. 5.13) for the times per item according to the order and also performed a statistical analysis (GLM–RM, see Sec. 3.3.3) for each order series. The result is unexpected, as all first orders were executed significantly faster than the third ones. This probably means that the design of the first order was structured more easily, as for all visualizations the same three orders were used. On the other hand, this means at least that the users did not have problems becoming comfortable with the system. We linked this effect back to the introduction (try and ask phase) we gave to our test participants each time before testing another visualization.

For each subject a mean time was calculated over all three orders per visualization. This results in the following average picking times per item: 4.2467s (StdDev: 1.101s)

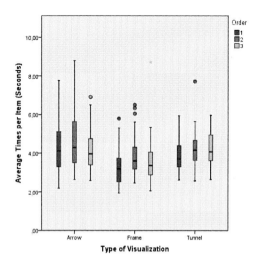

Figure 5.13.: Times (according to the order) in the second experiment.

for the Arrow, 3.581s (StdDev: 1.022s) for the Frame and 4.0812s (StdDev: 0.835s) for the Tunnel (see Fig. 5.14). The null hypotheses $H_{0-Time-Exp2}$ can be rejected and $H_{1-Time-Exp2}$ accepted as there exist significant differences between the mean values (GLM–RM, $p = 0.0 < 0.05 = \alpha$). The post hoc analysis shows the following results. Test participantss using the Frame are significantly faster than using the Arrow (PH–LSD, $p = 0.0 < 0.05 = \alpha$) or the Tunnel (PH–LSD, $p = 0.0 < 0.05 = \alpha$). The average difference between Frame and Arrow is 0.67s (StdDev: 0.161s) and between Frame and Tunnel 0.5s (StdDev: 0.119s) There was no significant difference between Arrow and Tunnel measured (PH–LSD, $p = 0.285 > 0.05 = \alpha$).

Subjective Observations

In addition to the objective measurements we will report on the observations we made during the experiment. We were able to separate the test participants exposed to the Arrow visualization into two groups. Test participants of the one group moved their head around the Arrow until they were sure about the correct box. They even got down on their knees to be on the same height with the arrow. While (time consumingly) examining the arrow from the top, the back and the side one can figure out the correct position. Thus in most cases, they picked from the correct box. However the latter is not the intention of an intuitive Augmented Reality visualization. The other group of

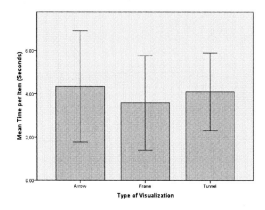

Figure 5.14.: The mean time to pick an item for each visualization, measured over all three batches. The error bar displays the double of the StdDev. The time for the Frame differs significantly from the Arrow and the Tunnel.

test participants just immediately picked the item, without elaborately examining the position of the Arrow. Naturally such users were faster, but made more mistakes.

Additionally we observed some people closing their eye which was not covered by the HMD. Those test participants felt uncomfortable and complained about cluttering of their vision. After telling them to open both eyes they felt much better and were able to perform the test.

Conclusions 1: Picking Test

In addition to showing that the test participants were significantly faster using the Frame, we determined that they made more mistakes using the Arrow than using the Frame or the Tunnel. This measurement is even more important as we do not have the benefit of a fast visualization if it is not error free.

We used the same Arrow visualization but with even smaller shelves than we used in the first experiment (Sec. 5.1), even so the test participants made fewer mistakes. We achieved this better result just by switching from the Sony Glastron to the Nomad Microvision HMD, but we are not yet sure of the exact reason. It could be either the better see-through capabilities or that the new display does not hide the rest of the field of view and provides much more peripheral vision than the Sony does. However the participants still made mistakes using the Arrow. The tunnel worked much better, but people still made mistakes, whereas the frame worked without mistakes.

In interviews, many users argued that they preferred the Tunnel to the Compass-like Arrow Meta Visualization. Indeed, the Compass-like Arrow indicates the direction but does not give an indication about the distance to the object and how far the user has to turn. The Tunnel instead intuitively shows by the strength of its bending how far and fast the user has to turn. Furthermore it immediately gives feedback about getting closer or further away. Yet, the test participants mentioned the Tunnel to be helpful for the coarse navigation but not for the fine navigation as it is sometimes ambiguous as to which box it points to. For the fine navigation they preferred the frame, because it was more exact. This relates directly to the fact that people did not make mistakes using the frame.

Personality and Performance Tests

To answer the second question (whether we can predict efficient users of Augmented Reality systems) we make use of statistical correlation and regression. This analysis is based on the advice of [7, 133, 15].

First of all we calculated correlations between all parameters collected in phase A of the test, using the methods described in Sec5.2.2 with the overall picking task performance. This resulted in the two tables (Fig. 5.15 and Fig. 5.16). The first value, the *r-value* indicates strength and direction (positive/negative) of the correlation, where closer to (-)1 is better. A significant correlation is marked with a "*". The second line indicates the *p-value*, the probability of error of the r-value. This means that the r-value is significant if the p-value falls below the α niveau of 5%. The last line describes the number of samples used.

The two tables (Fig. 5.15 and Fig. 5.16) show significant correlations for Neuroticism, Openness for experiences, the Ego Shooter test and the 3D Maze test. All correlations are positive, because of the positive r-values, which means if one value increases, the other does as well.

However, it is important to make sure that these predictors are independent of each other. Otherwise the probabilities estimated with the regression analysis will not be correct because of redundancies. We calculated a correlation between these four predictors. The results of this correlation (Fig. 5.17) show a problem: Neuroticism correlates with Openness and Ego Shooter correlates with 3D Maze. This means a high value in Neuroticism goes along with a high value in Openness The problem with a possible intercorrelation of the Big Five was already reported by Haupt [55]. The correlation between Ego Shooter and 3D Maze was not unexpected, as our aim of both tests was to assess the ability of using 3D computer games without having an idea, which is better suited. To this end we had to sort out these redundant predictors. In each case we did not sort out the predictor, which has a stronger bivariate correlation (compare Fig. 5.15 and Fig. 5.16) with the picking performance. That means we finally used Ego Shooter and Neuroticism as independent variables for the regression analysis.

Correlations

		Picking (Average Time)
Neuroticism	Pearson Correlation	.455**
	Sig. (2-tailed)	.007
	N	34
Extraversion	Pearson Correlation	.085
	Sig. (2-tailed)	.634
	N	34
Openness	Pearson Correlation	.396*
	Sig. (2-tailed)	.020
	N	34
Agreeableness	Pearson Correlation	.259
	Sig. (2-tailed)	.139
	N	34
Conscientiousness	Pearson Correlation	-.159
	Sig. (2-tailed)	.369
	N	34
Verbal IQ	Pearson Correlation	.009
	Sig. (2-tailed)	.958
	N	34
Numeric IQ	Pearson Correlation	-.145
	Sig. (2-tailed)	.413
	N	34
Figural IQ	Pearson Correlation	-.016
	Sig. (2-tailed)	.927
	N	34
Deductive IQ	Pearson Correlation	.000
	Sig. (2-tailed)	.998
	N	34

**. Correlation is significant at the 0.01 level (2-tailed).
*. Correlation is significant at the 0.05 level (2-tailed).

Figure 5.15.: Correlation of possible predictors with the measured picking performances (table 1 of 2). The first five rows show the NEO-PI-R values and the last four rows show the Structured IQ values.

Correlations

		Picking (Average Time)
Ego Shooter (Average Time)	Pearson Correlation	.611**
	Sig. (2-tailed)	.000
	N	34
3D Maze (Average Time)	Pearson Correlation	.556**
	Sig. (2-tailed)	.001
	N	34
Frequent Computer Gamer (past)	Pearson Correlation	.177
	Sig. (2-tailed)	.316
	N	34
Frequent Computer Gamer (present)	Pearson Correlation	.216
	Sig. (2-tailed)	.220
	N	34
Type of Computer Games (3D)	Pearson Correlation	.192
	Sig. (2-tailed)	.276
	N	34
Computer (Experience)	Pearson Correlation	.085
	Sig. (2-tailed)	.632
	N	34
HMD (Experience)	Pearson Correlation	−.085
	Sig. (2-tailed)	.631
	N	34
PDA (Experience)	Pearson Correlation	.174
	Sig. (2-tailed)	.324
	N	34
Mobile Phone (Experience)	Pearson Correlation	−.013
	Sig. (2-tailed)	.944
	N	34
VR (Experience)	Pearson Correlation	−.209
	Sig. (2-tailed)	.235
	N	34
CAD (Experience)	Pearson Correlation	−.117
	Sig. (2-tailed)	.509
	N	34

**. Correlation is significant at the 0.01 level (2-tailed).

Figure 5.16.: Correlation of possible predictors with the measured picking performances (table 2 of 2). The first two lines show the correlation with the times for the 3D Gaming tests. The rest shows the correlation with the answers from the questionnaire.

Correlations

		Neuroticism	Openness	Ego Shooter (Average Time)	3D Maze (Average Time)
Neuroticism	Pearson Correlation	1	.450**	.078	.102
	Sig. (2-tailed)		.008	.660	.565
	N	34	34	34	34
Openness	Pearson Correlation	.450**	1	.218	.094
	Sig. (2-tailed)	.008		.216	.598
	N	34	34	34	34
Ego Shooter (Average Time)	Pearson Correlation	.078	.218	1	.693**
	Sig. (2-tailed)	.660	.216		.000
	N	34	34	34	34
3D Maze (Average Time)	Pearson Correlation	.102	.094	.693**	1
	Sig. (2-tailed)	.565	.598	.000	
	N	34	34	34	34

**. Correlation is significant at the 0.01 level (2-tailed).

Figure 5.17.: Correlation of the significant predictors.

Choosing these two predictors, the regression analysis delivered two models. The first model only uses Ego Shooter as a predictor and can explain $R^2 = 37.3\%$ of the variance in Picking Performance of the sample. This (R^2) has to be adjusted (depending on sample size and number of predictors), which reduces the amount of explainable variance to $R^2 = 35.3\%$. The second model uses Ego Shooter and Neuroticism as predictors with $R^2 = 54.1\%$ and adjusted $R^2 = 51.0\%$. Hence, the second model seems to be a better predictor.

Fig. 5.18 reveals a deeper look into the two models. The displayed B values make it possible to set up the equation describing the relationship between the test participants' results in the Ego Shooter and Neuroticism tests and their possible result with the picking systems. This would look like: $PickingPerformance = -0.151 + 0.289 * EgoShooter + 0.004 * Neuroticism$. But as the Constant (-0,151) is not significantly estimated, we have to be careful with this equation. However, what gives a better view into the results of this test are the Standardized Beta Coefficients (which brings the B-values on a common scale). For the Ego Shooter the adjusted Beta is 0.579 and for Neuroticism it is 0.410, which shows that both values have an influence on the picking performance in the same order of magnitude.

Conclusions 2: Personality and Performance Tests

What is the meaning of this result for the identification of likely successful users? We only conducted an exploratory study, with a small sample. This means we can not yet draw a conclusion but rather we are able to set up the hypothesis that the Ego Shooter Test and the Neuroticism can predict the performance of a human working with Augmented Reality (Picking) systems: The longer the Ego Shooter time (respectively the higher the

Model		Unstandardized Coefficients		Standardized Coefficients		
		B	Std. Error	Beta	t	Sig.
1	(Constant)	.112	.292		.383	.705
	Ego Shooter (Average Time)	.305	.070	.611	4.361	.000
2	(Constant)	-.151	.266		-.568	.574
	Ego Shooter (Average Time)	.289	.061	.579	4.733	.000
	Neuroticism	.004	.001	.410	3.353	.002

Figure 5.18.: Two regression models of the strong and independent predictors with the average picking task performance time.

value for Neuroticism) the worse the Picking performance. To get evidence for this, a further experiment is required to statistically prove/reject this hypothesis. This probably needs a much bigger sample and should additionally focus on the longterm behavior with AR systems to fully exclude learning effects.

What furthermore remains are two practical facts we probably should consider when performing user evaluations with new technologies like Augmented Reality. a) Someone who is not familiar with computers or virtual environments will probably have problems using an Augmented Reality system. Novice computer users have to be elaborately trained (to be able to use Augmented Reality), before the actual test can be performed. b) Users with a high value in Neuroticism (afraid, bilious, depressive, social shyness, impulsive, sensitive) performed worse with our AR system. An explanation could be the following: imagine someone who has to test an imperfect new technology but has a neurotic personality. In this case, you would not simply go there and put on the HMD (something you have never seen before), immediately asking about something you do not understand ad-hoc and confidently hurdling all the minor usability problems – on the contrary you would probably be careful and shy in exploring this new technology. If this explanation is right, we can furthermore hypothesize that neurotic people just need more time to learn to handle new technologies. That means we have to handle such persons even more carefully when they enter our lab and support them in getting really comfortable with Augmented Reality.

5.2.4. Conclusions

The results and conclusion of our second picking experiment will be shortly summarized below.

Visualization

Users were fastest and made no errors when using the Frame visualization supported by the Compass-like Meta Visualization. Nevertheless, the users complained that the Compass-like arrow, indeed gava a clear directional information, but did not provide information on the distance. This distance information was provided by the Tunnel, which was thus preferred as a Meta Visualization over the Compass-like Arrow. However, the Tunnel was not a good Box Finding visualization.

Evaluating AR / Lessons Learned

An elaborate introduction like our try-and-ask phase is required in order not to confound the results of a user-based Augmented Reality interface evaluation. Making use of this intensive introduction, we did not encounter huge learning effects as in the first experiment (5.1). We furthermore hypothesize that the learning curve is influenced by the personality type (neurotic) and the experience with virtual technologies. Depending on the type of user, the introduction has to be extended (sometimes as long or longer than the actual test time).

System

By switching from the Sony HMD to the Nomad HMD, keeping the same visualization, users made less errors, even with smaller boxes.

5.3. First Public Evaluation

From the results of the previous evaluation, we tried to combine the advantages of the visualizations by combining the Tunnel with the Frame (and removing the Compass-like Arrow). This provides a good Meta Visualization to guide the user's view to the actual augmentation by using the Tunnel and an exact Box Finding visualization by the Frame. This combined visualization can be seen in Fig. 5.19.

We presented this system again at an exhibition, where more than 30 visitors tried it. After a very short instruction almost all people were able to easily handle our visualization. We just explained: "There is a tunnel starting in front of your face and ending at the box from which you have to pick the candy." People did not make mistakes but some of them complained about being disturbed by the tunnel when they wanted to finally pick the item. We observed again that some people were moving their heads around the box to examine whether they were seeing the visualization in front of the correct box - an effect we already noticed by observing people using the Arrow to highlight the box. However this time, users who did not move their head around the box did not pick wrongly. Nevertheless, some users had problems identifying the Frame behind the Tunnel. The problem was, that we simply cluttered the visualization too much.

Figure 5.19.: Two perspektivs of the opaque Square-based Tunnel in combination with the Frame. Picture a) shows only the Meta Visualization, when the Box Finding visualization is not in the field of view. In b) one sees the cluttering. The Tunnel occludes the Frame.

5.4. Improving the Meta Visualizations

With the experiences of our previous experiments and some new ideas in mind, we designed a new experiment. An overview of this experiment is given in Fig. 5.20.

Objectives		Speeding up the process by improving the meta visualization (reducing clutter) and proving the Frame as perfect Box Finding visualization
Experimental Setup	Hypothesis	A) Participants will not make errors using the Frame b)There will differences in the task performance dependent on the Meta Visualization used.
	Independent Variables	Type of Meta Visualization : opaque Square-based Tunnel, semi-transparent Square-based Tunnel, semi-transparent Ring-based Tunnel, Compass-like Arrow with Frame
	Dependent Variables	Time, Failure
	Test Method	Formal experiment, within-subject design, warehouse 2 (96 small boxes on one wall), manual user input (click-turn-wheel)
Results	AR - Visualization	a) No errors made b) Ring-based Tunnel good for smooth movements c) Compass-like Arrow good for direction
	System	Click-turn-wheel makes people slower (than Wizard-of-Oz), but is required.
	Evaluating AR / Lessons Learned	The elaborate try-and-ask phase was shown again to compensate for learning effects.

Figure 5.20.: Overview of experiment to improve the Meta Visualisation

5.4.1. Objectives

The main intention of this experiment was to speed up the picking process by reducing the cluttering of the Tunnel. We tried to achieve this by fading the Tunnel to transparent when the actual Box Finding visualization was in the users' field of view. Furthermore we simply aimed at exploring alternative Meta Visualizations.

5.4.2. Experimental Setup

The experimental hardware setup consisted of the components which were already used in the previous test series: The Nomad-HMD and the shelves consisting of 96 boxes.

The only difference was to attach the Powermate from Griffin Technology as an input device (see Sec. 4.2.1). This device was placed at a fixed position in front of the shelf and replaced our previous Wizard of Oz-based method to step through the picking list.

Independent Variable: Type of Meta Visualization at 4 Levels

In the previous experiment (Sec. 5.2) and during the exhibition (Sec. 5.3) people had not picked from a wrong box when the box was highlighted by a Frame (like the one in Fig. 5.9), whereas the 3D Arrow (Fig. 5.8) and the Tunnel (Fig. 5.10) alone were no safe indicators for the right box. Due to this fact, we based all further developments on the Frame, exploring how to combine it with different Meta Visualizations to initially guide users toward having the Box Finding visualization within their field of view. We designed four Meta Visualizations to go with the Frame.

Opaque Square-based Tunnel with Frame As first navigation we chose the square-based Tunnel in combination with the Frame (Fig. 5.19) as it was presented by us at the public exhibition (section 5.3). This visualization metaphor was part of our evaluation even though we already knew about the drawbacks of cluttering. We wanted to see how it performed in comparison to the other visualizations.

Semi-transparent Square-based Tunnel with Frame To compensate for the disadvantage of occluding the Frame by the opaque square-based Tunnel we slightly modified the latter visualization: The purpose of the Tunnel is to bring the next relevant augmentation to the field of view. That means guiding the user's gaze until he has found the Frame augmentation in front of the box. After the user has found the Frame he does not need the Tunnel anymore. So we can simply fade it out when the Frame is in the user's field of view. As Biocca et al[14] proposed, the fading is done according to the dot product between the vector of the user's view and the vector pointing out of the box. This visualization can be seen in Fig. 5.21.

Compass-like Arrow with Frame As we found the Frame in combination with the Meta Visualization (consisting of the Compass-like Arrow and a rubber band) (Fig. 5.9) in the previous series to be good (but not perfect), we used it in this evaluation as a measure of ground truth.

Semi-transparent Ring-based Tunnel with Frame As we had had some problems with artifacts and the overlay of several semi-transparent square corners when fading the Tunnel out, we replaced the squares of the Tunnel by rings. So we designed a Tunnel with rings in combination with the Frame (see Fig. 5.22). The rings were fading to transparent on the basis of the same function we already used for the semi-transparent square-based Tunnel.

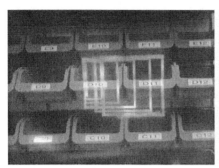

Figure 5.21.: The semi-transparent Square-based Tunnel. The Tunnel fades out when the real visualization is in the field of view.

Figure 5.22.: The semi-transparent Ring-Tunnel. The tunnel fades out when the real visualization is in the field of view.

Hypotheses

The intention of this experiment was similar to the last experiment. We were looking for the fastest visualization which also supported an error free picking. As the visualizations using the Frame from the last experiment were shown to be error proof and in this experiment all our visualizations used the Frame, we set up the hypothesis that people would not make a single error using our visualization.

$H_{1-Error-Exp4}$: No user will make an error when using our different visualizations.

The main measurable difference we expected to find was in the task performance time. This leads to the following hypothesis:

$H_{1-Time-Exp4}$: There will be a difference in the task completion times according to the type of visualization.

We could have used the experiences of the previous evaluations to set up more specific hypotheses, that, for example, the opaque Tunnel would be slower than the semi-transparent Tunnel. The probability to prove such a directed hypothesis is much higher than proving an undirected one. As we were looking for results, rather than looking for easy to reject hypotheses, we decided not to examine such theories further.

Test Method and Dependent Variables

We basically used the same test setup as in the second experiment (Sec. 5.2). The single independent variable was the type of Meta Visualization. The four levels of our variable were the opaque Square based Tunnel, the semi-transparent Square-based Tunnel, the Compass-like Arrow and the semi-transparent Ring-based Tunnel. This time we decided to involve less test participants, but to spend more time observing and interviewing the users. Thus each user had to spent about 45 minutes on the test.

Once again we solely tested the picking task. Before starting with each visualization there was an elaborate introduction for each visualization (try-and-ask-phase). This introduction is, as we learned from the earlier experiments, a good way to compensate for the confusing/learning effect resulting in long picking times in the first order. We used the same order structure as we did in the second experiment (Sec. 5.2): Each test participant had to execute 3 orders with 9 items for each of the 4 visualizations (within-subject design). The order in which the test participants had to use these visualizations was permuted to compensate for learning effects.

The test participants this time controlled the work flow themselves by pressing a button, placed in front of the shelf. Again we provided auditory feedback of the state change, by playing a sound when the button was pressed. As dependent variables, we logged the times when people pressed the button and recorded the number of wrongly picked items.

5.4.3. Results

We used 14 test participants between 20 and 50 years (mean age: 27,7, StdDev: 7,5). Each of them again received some candies as reimbursement.

Errors

14 of the the 14 measurement series for each type of visualization were valid. During all of the 1512 picks none of the users picked a wrong item. Ten times the button was pressed twice, which produced missing order lines. In consequence, we can accept $H_{1-Error-Exp4}$, which states that people would not make any errors with any of our visualizations. We hypothesized this due to the fact that they did not make any with the Frame in the second test. Even so, we were still surprised that people really did not make a mistake using any Meta Visualization in combination with the Frame.

Picking Times

As for the errors, we had 14 of the 14 measurement series for each type of visualization, leaving out the ten times when the button was pressed twice. We conducted some pre-evaluations of the data to check for learning effects between the three orders for each visualization. To this end, we firstly drew a box plot (Fig. 5.23.) for the times per item according to the order. Secondly, we performed a statistical analysis (GLM–RM, see Sec. 3.3.3) for each order series. This analysis shows that the users using the semi-transparent Square-based Tunnel were faster in the first order than in the second. We do not have an explanation for that. Furthermore it shows that participants using the opaque Ring-based Tunnel were faster with the third order than with the first and second. Hence the opaque Ring-base Tunnel is the only visualization showing significant learning effects in this short experiment.

To treat all orders the same, the mean values of the times were averaged over all three orders of a visualization for each user. From this, the mean values for each type of visualization were calculated (see Fig. 5.24): opaque S-Tunnel 6.57s (StDev: 0.78s), semi-transparent S-Tunnel 6.32s (StDev: 0.76s), Compass-like Arrow 5.93s (StDev: 0.89s) and semi-transparent R-Tunnel 6,04s (StDev: 0,69s). The null hypotheses $H_{0-Time-Exp4}$ can be rejected and $H_{1-Time-Exp4}$ can be accepted, as there exist significant differences between the mean values (GLM-RM, $p = 0.033 < 0.05 = \alpha$). The post hoc tests show where exactly the differences are. The Compass-like Arrow is, on average, 0.65s (StDev:0.19s) faster than the opaque Square-based Tunnel (PH–LSD,$p = 0.005 < 0.05 = \alpha$). The semi-transparent Ring-based Tunnel is on average 0.53s (StDev:0.23s) faster than the opaque Square-based Tunnel (PH–LSD, $p = 0.005 < 0.037 = \alpha$). All other pairwise tests are not significant.

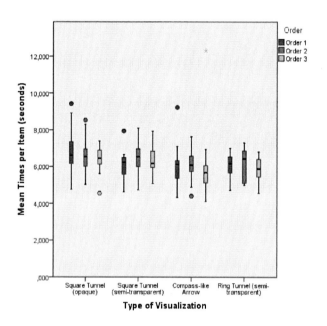

Figure 5.23.: Times (per order) in the third experiment. Outliers are marked separately.

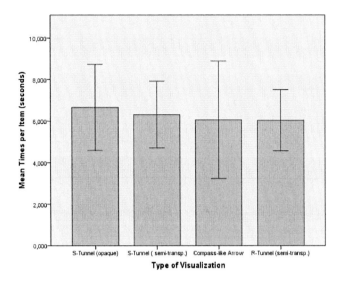

Figure 5.24.: The mean time to pick an item for each visualization, measured over all three batches. The error bar displays the double of the StdDev. The opaque Square-based Tunnel is significantly slower then the the semi-transparent Ring-based Tunnel and the Compass-like Arrow.

Subjective Observations

In this section we will discuss the subjective observations we made during the experiment and the subsequent interviews. Most of these facts coincide with the previously discussed objective results, or are at least not contradictory.

The users complained about the opaque Square-based Tunnel as it cluttered the view and they had to look around it to see the Frame. For that reason it was determined to be the worst solution. This fact is directly linked to the significant speed difference compared to the other visualizations. The semi-transparent S-Tunnel was determined to be the second worst visualization.

The subjectively best solutions were the semi-transparent Ring-based Tunnel and the Compass-like Arrow. Again this correlates with the objectively measured facts. A clear best solution was indeterminable due to several arguments. Some people preferred the Compass-like Arrow over the opaque Ring-based Tunnel, due to less cluttering. At first sight, this is a confusing result as the tunnel is faded semi-transparently and should not clutter the Frame. However we used the proposed solution of Biocca et al [14] to fade the Tunnel according to the dot-product of the start and end vector of the Bezier curve. This works fine, if one looks straight at the box. But if one looks, for example, from above at a box which is quite low, the Frame is in the center of the HMD, but the transparency function does not work properly and does not fade out. On account of this, users liked the opaque Ring-based Tunnel more than the Square-based Tunnels. They could distinguish more easily the Rings from the Frame.

We received several interesting remarks regarding the difference between the Tunnel and the Compass-like Arrow. Users liked the Compass-like Arrow, because it directly indicated where to look. Nevertheless it did not provide any distance information. So people sometimes moved their heads too fast and looked past the Frame and then had to move their head back. In comparison the Tunnel is not a good indicator for the direction at the very first beginning of the movement, but it later provides better information about the remaining distance to the target. Furthermore people argued that sometimes they had to take one step back to see more elements of the tunnel. Thus the Tunnel was not a good indicator if one stayed directly (about 0.5m) in front of the shelf. One user considered the Tunnel to be more ergonomic, on the basis use of the Arrow forced users to make robot like movements.

5.4.4. Conclusions

Here we will shortly summarize the achievements of this third formal experiment.

Visualization

In this iteration of user studies we have been able to prove the method of displaying the Frame as a perfect Box Finding visualization. We have succeeded in designing a reliable system to precisely highlight boxes for the order picking process – something we would not have thought after the numerous errors people made in our first experiment (Sec. 5.1), when we displayed an Arrow in the Sony Glasstron HMD.

After finding a good Box Finding visualization, the Meta Navigation is still questionable. We basically compared a Compass-like Arrow in combination with a rubber band against a Bezier curve-based tunnel to guide the users' attention to off-screen visualizations. Our best solution so far is that the Tunnel has to behave in an attentive way in order not to clutter the display: it has to disappear (fading to transparent), when it is currently not needed. The idea to fade the Tunnel to semi-transparent when the target is in the field of view was proved to be good, but the way we implemented this was not. Fading on the basis of the dot-product between start and end tangent of the Bezier curve does not work in some cases. On account of this, there was need for improvement.

However the semi-transparent Ring-based Tunnel already performed (using the bad fading function) as well as the Compass-like Arrow, due to the fact that users could distinguish between the elements of the Tunnel and the Frame due to the geometrical difference and not the effect of transparency. The Compass-like Arrow was good for giving direction information – in particular when staying directly in front of the shelf. The Tunnel, on the other hand, gave good feedback about the distance. Altogether, it was more valuable when staying a step back from the shelf.

Evaluating AR/ Lessons Learned

As a side effect, while executing this experiment we could again see the benefit of our try-and-ask phase at the beginning of each experiment. Each participant got an introduction with some items to pick in advance. Participants were allowed to try the new visualizations until they understood them. Applying this technique, we could compensate for huge learning and fascination effects while executing several orders in a row.

Furthermore we can summarize that the informal feedback we got by observing the people at the exhibitions was at least as valuable as the feedback from the formal user studies.

System

The overall average Picking time per item increased by about 2 seconds compared to the previous experiment in Sec. 5.2. The only thing we changed between the two setups was the input method (Wizard-of-Oz to manual user input by the Button - see Sec. 4.2.1). This change was indispensable, as the next step in the evolution of the Pick-by-Vision (AR) system will be a completely mobile setup in a realistic scenario.

5.5. Bringing AR into a Real Warehouse - A Pretest

To get a clearer understanding of Pick-by-Vision (AR) and the difference between the users' performance using the Tunnel and the Compass-like Arrow as Meta Visualizations, we moved our setup from the single shelf (Fig. 5.7b) to a realistic warehouse-like setting, consisting of four shelves and two aisles (see Fig. 5.25). In the previous setup people basically had to find boxes in a single plane: they mostly had to turn their head by no more than 90 degrees to see the target.

In our new setup, we exposed people to a situation in which they had to navigate between several shelves to find the boxes, and the boxes could be rather high above them, low at their feet, very far to the left or right, or also on a shelf behind them. Meta Visualizations thus had to cover nearly the entire range of 4π steradians. Furthermore, test participants had to walk from box to box instead of having everything just about reachable at arm's length. The distances for the augmentations varied from relatively close distances to about four meters. This was a challenge to the depth perception that needs to be provided by the visualization.

Due to external constraints (including a limited time frame), we had to use our new setup to perform summative evaluations comparing our Pick-by-Vision (AR) system with a traditional paper-based list system, as well as with the Pick-by-Vision (2D) System. This constraint was accompanied by the fact that we were going to use our system for the first time under realistic conditions. We knew that our Pick-by-Vision (AR) system was far from perfect, and we had a countless number of variations in what the visualization could look like and how the system could behave. That is why we had to find an efficient way of optimizing our AR System for this new warehouse.

5.5.1. Informal Evaluation and Adaption to the New Warehouse

After several discussions, we changed our approach toward evaluating user interfaces. We moved from the quantitative approach using rigid usability studies to a qualitative approach. With the new qualitative approach, we heavily reduced the number of test participants. However, we followed them through every move and turn they made. We watched all interactions and all facial or body reactions extensively and discussed their reactions with them during and after the test.

To this end, we observed 8 users in-depth, each of them for about 2 hours. Our test participants were Augmented Reality experts, operative workers, or students. During the test we presented the different visualizations to them and reacted to their comments immediately by changing the visualization on an ad-hoc basis. This was possible because the parameters of the visualization were designed to be changeable on the fly (compare Sec. 4.1). This was extremely helpful since we could discuss the changed approach immediately with the test persons. In retrospect, this change of evaluation policy was one of the best decisions we made in the entire process.

Figure 5.25.: The warehouse with 4 shelves. The test user wears the Nomad HMD from
the previous experiments. The WiFi-connected wireless PC is carried in a
small backpack. As system control, we mounted the game show-like buzzer
on the user's belt. The tracking is done via the red encircled infrared-based
optical camera system.

The main results of these informal pre-evaluations were that most people preferred the Tunnel over the Compass-like Arrow as Meta Visualization. This was, among other things, because the test users expressed similar positive preferences for the Tunnel as users did in previous tests. For example the Tunnel provides continuous information on the distance to the box. Furthermore, it gives additional depth perception for the Box Finding visualisation, especially for boxes close to the floor. Our most successful improvement for the Tunnel (with the most impact at the lowest costs) was to reduce the thickness of the rings of the Tunnel to a minimum. This way, we minimized the cluttering by the rings and made the Tunnel easily distinguishable from the Frame (see Fig. 5.26).

This was not the only problem we could solve by minimizing the thickness of the rings. In the last experiment in Sec. 5.4, we showed the possible benefit of fading out the Tunnel when the actual Box Finding visualization is in the field of view. However we realized that the users have a benefit of the Tunnel in the larger warehouse, even when the Box Finding visualization is in the field of view: it provides quite important depth queues in the larger distances. For that reason, we decided not to fade out the Tunnel, when the Box Finding visualization is in the field of view. This is only possible because of the reduced thickness of the rings. This depth queue benefit of the Tunnel can be seen in Fig. 5.26.

The reduced thickness makes the Tunnel unobtrusive. One thus can simply see through it. This fact is quite important, as the worker still has to interact with his real environment. On the one hand we have to give the worker a free view, so that he does not stumble. On the other hand we can not decide whether he currently needs the visualization or not. For example we can not decide whether the worker has just picked an item and has to look at the item to compare the article number (compare Fig. 5.27) or whether the worker is still looking for the box to pick from. Both problems are addressed by the thin Tunnel.

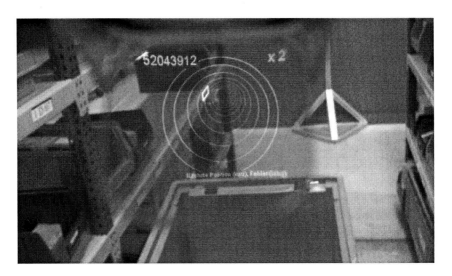

Figure 5.26.: A photograph through a tracked head-mounted display (HMD - compare
Fig. 5.7a). The visualization shows a Tunnel twisting to the left, with a
square Frame at the end highlighting the label under a box in a warehouse
from which a worker has to pick items.

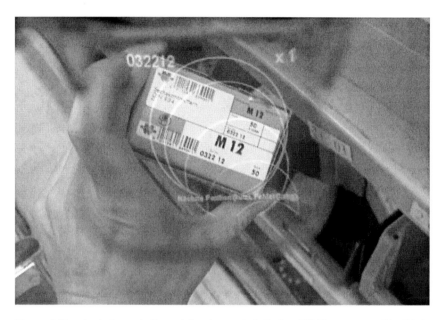

Figure 5.27.: A photograph through head-mounted display (HMD - compare Fig. 5.7a). The visualization shows that users can see through the Tunnel, to read, for example, the number printed on an article.

5.6. First Comparative User Study in a Real Warehouse

After the first fine-tuning of our system in the real warehouse (Sec. 5.5) we wanted to the see the performance of our Pick-by-Vision (AR) system in the new setup and thus designed a new experiment. A brief overview is given in Fig. 5.28

Objectives		Evaluating the performance of Pick-by-Vison (AR), Pick-by-Vision (2D) compared to Pick-by-Paper in a real warehouse.
Experimental Setup	Hypothesis	Users make less errors and are faster with Pick-by-Vision (AR) than with Pick-by-Vision (2D), which is again better than Pick-by-Paper. Pick-by-Vision produces more strain
	Independent Variables	Type of Picking System: Pick-by-Vision (AR), Pick-by-Vision (2D), Pick-by-Paper
	Dependent Variables	Time, Failure, (subjective) Strain
	Test Method	Formal Experiment, within-subject design, warehouse 3: a real warehouse with 2 aisles and real items, manual user input :click-turn-wheel
Results	AR - Visualization	No one picked out of the wrong box. Tunnel only works well, if the user is not turned away from the box further than 80 degrees.
	System	Several usability problems found (no step back, jumped over items), no increased strain
	Evaluating AR / Lessons Learned	It is quite important to go to the real environment for testing the system and to elicit the the real requirements.

Figure 5.28.: Overview of first comparative user study in a real warehouse.

5.6.1. Objectives

The main intention of this study was to get a first impression whether we will have a benefit of using a Pick-by-Vision (AR) system over a Pick-by-Vision(2D) system. As a ground truth we choose a conventional Pick-by-Paper System, which makes use of printed lists such as the one in Fig. 2.1.

5.6.2. Experimental Setup

For our testing environment, we used the warehouse described in Sec. 5.5 and Fig. 5.25. The test participants were equipped with the Nomad HMD and a light-weight laptop carried in a backpack. The click-turn-wheel input device was mounted to the wrist as described in Sec. 4.2.1. The only interaction required was to click the click-turn-wheel, to request the next order and to switch to the next order line.

Independent Variable: Type of Picking System at 3 Levels

In this test we used three different picking technologies. Beside the Pick-by-Paper System, we used our Pick-by-Vision System with two different configurations AR and 2D (compare Sec. 2.3). As the Pick-by-Vision (AR) System is under continuous development, we started to identify each iteration. To this end, this version is 1.0 as it is the first one we use in a real environment - all development stages are described in Sec. 6.1.

Test Method & Dependent Variables

The main dependent variables we discuss here are picking performance (time), errors and strain. To analyze the latter variable (strain), we used the common NASA Task Load Index (TLX) test [53], which is based on a subjective questionnaire.

The experimental design was of type within-subject: Each test participant had to carry out the test with each of the three technologies, with the start sequence of the technologies being permuted between test participants. The test participants had to fulfill six orders with all in all 30 order lines (Ø5.0 order lines per order) and 61 items (Ø2.03 items per order line) for each technology. To exclude the risk of learning effects the orders were different between the three picking technologies. This is in contrast to the previous experiments, but this time the shelves and items are not identical as in the previous setups. Within one order the order lines are different but the total number of items, their weight, and the ranges to go and to pick were the same. To compensate for the learning effects, when using an Augmented Reality system for the first time, we made use of an extensive try-and-ask phase (as in the previous experiments).

Hypotheses

We set up the following hypotheses, while keeping in mind that our Pick-by-Vision systems are far from being perfect and it is actually too early to test the system performance in a summative experiment (compare Sec. 3.2.2). For that reason, these hypotheses should rather be seen as research questions, which reflect our focus. The mathematical proof of these hypotheses is a minor intention. In the previous tests we improved our Pick-by-Vision (AR) System to support an efficient and error free-picking [124], and for the Pick-by-Vision (2D) system we knew that it performs better than a paper based system [115]. For that reason we set up the hypothesis that the use of a more advanced picking system results in a better performance:

$H_{1-Error-Exp6}$: Users make less errors using a Pick-by-Vision (AR) system than by using a Pick-by-Vision (2D) system, which again produces fewer errors than using Pick-by-Paper.

$H_{1-Time-Exp6}$: Users are faster using a Pick-by-Vision (AR) system than by using a Pick-by-Vision (2D) system, which again supports a faster picking than using Pick-by-Paper.

Beside improving the logistic performance figures, one design goal is not to produce stress using the Pick-by-Vision systems. Nevertheless, especially because of the weight of the hardware and the problem of switching between real and virtual focal planes (compare Sec. 4.2.2), we assume that using the Pick-by-Vision systems results in larger strain than working with Pick-by-Paper. As measure of strain we mainly use the NASA TLX test. Furthermore, we used three positive questions (Likert scale) to gain insight into the general comfort when wearing the Pick-by-Vision system, as well as whether or not the users felt constrained by the HMD or the visualization. So our third hypothesis is:

$H_{1-Strain-Exp6}$: Picking supported by Pick-by-Vision systems produces more strain than picking with Pick-by-Paper.

5.6.3. Results

We had 19 test participants (16 male/3 female) between the age of 18 and 45 (mean age: 27.2, StdDev: 6.78). The test participants were mainly individuals from all areas of the university, friends, as well as three professional order picking workers. As we said at the beginning, we compare our still imperfect Pick-by-Vision System (here at Version 1.0) with an established technology. For that reason we expect to find several unlucky factors, which could lead to a generally bad performance of our Pick-by-Vision systems in comparison to Pick-by-Paper. By an in-depth observation of the experiment, we can prevent such factors from confounding the results of the experiment.

Error Rates On the first view, according to the error rate, test participants performed significantly worse using both Pick-by-Vision systems as compared to the Pick-by-Paper system. However, most errors can be traced back to a bad system configuration. Sometimes users accidently skipped over an order line and had no possibility to go back in the system. Hence, errors of this category recognized by the users were discounted. However, two missing order lines within Pick-by-Vision (AR) were not recognized by the test participants. After this correction, the most common error was picking the wrong amount. Neither Pick-by-Paper nor either of the Pick-by-Vision systems had a method to avoid this error. Furthermore, a wrong item was picked using Pick-by-Paper and also by using the Pick-by-Vision (2D) system, but no wrong items were picked using the Pick-by-Vision (AR) system. As there were only a few errors made at all, we leave out the statistical analysis and look at the total number of wrongly picked items, which is shown in Fig. 5.29. After correction, 4 errors were made with Pick-by-Vision (AR), 6 with Pick-by-Vision (2D) and 8 with Pick-by-Paper.

The exploration of the errors shows that there is a tendency for a better performance of the Pick-by-Vision systems and thus for an acceptance of $H_{1-Error-Exp6}$. However, the sample is too small to make a reliable assertion of the hypothesis. More importantly, these results show that a lot of usability problems remain in the Pick-by-Vision prototypes.

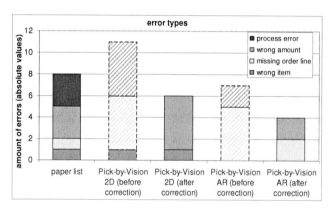

Figure 5.29.: Kinds of picking the overall errors for the three technologies before and after the correction of systematic errors.

Time   We had 19 valid measurement series for each picking technology. As the orders within one level of the dependent variable are different, there exists one summarized time measurement per user for each technology. The box plot for the times is shown in Fig. 5.30. In particular it shows that most users of the Pick-by-Vision systems are relatively fast, but that there are a few slow users, because the median is quite low. The average values are shown in Fig. 5.31 and they are as follows: 12,41 min (StdDev:2.3min) for Pick-by-Vision (AR), 12,65 min (StdDev: 3.08min) for Pick-by-Vision (2D) and 13,86 min (StdDev: 2.20min) for Pick-by-Paper. This difference is significant (GLM-RM, $p = 0.025 < 0.05 = \alpha$). Pick-by-Vision (AR) is on average 1.456min (StdDev: 0.483) faster than Pick-by-Paper (PH–LSD, $p = 0.007 < 0.05 = \alpha$). Pick-by-Vision (2D) is on average 1.211min (StDev: 0.487) faster thanPick-by-Paper (PH–LSD, $p = 0.023 < 0.05 = \alpha$). The other differences are not significant. The difference between Pick-by-Vision (AR) and Pick-by-Vision (2D) is not significant (PH–LSD, $p = 0.710 > 0.05 = \alpha$) Only because of the latter reason we cannot reject the null hypotheses and thus cannot accept $H_{1-Time-Exp6}$.

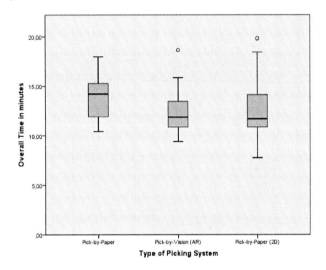

Figure 5.30.: Box plot of the times for experiment 6.

Subjective Measured Strain The NASA TLX test resulted in nearly the same values (25-28) for all systems. An analysis of the questionnaire did not show that users felt uncomfortable or constrained by using the Pick-by-Vision system. However we should be careful with accepting the null hypothesis ($H_{0-Strain-Exp6}$ – the use of a Pick-by-Vison system does not produce more strain), as the users had to work only twice for about a quarter of an hour with the system.

Other Observations The test was accompanied by a brief questionnaire. We could again find a tendency that people with previous experience with 3D user interfaces performed better in both logistics operating figures (faster with fewer errors).

By observing and interviewing the participants, we realized that the Tunnel was not yet optimal. It performed well when people were not turned away by more than about

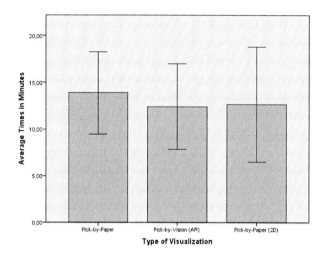

Figure 5.31.: Average order picking time for the three technologies

80 degrees from the box to pick from. For wider angles, the Tunnel was not able to indicate clearly in which direction the user had to start rotating. Furthermore, if the box was much further away than 100 degrees, the mathematically shortest path often went through the legs or over the head of the user. Unfortunately, this was not the fastest path for the user to turn towards the box.

5.6.4. Conclusions

This section presents the conclusion of our first experiment in a realistic warehouse.

Visualization

In this experiment, we again could show that the Frame visualization is a perfect Box Finding visualization. However, bringing the Augmented Reality visualization to a realistic setup also showed that the Tunnel is not yet perfect. It is only good in guiding the user to the object out of sight, when the user is not turned away for more then 80 degrees from the object. To this end, there are improvements required to the Tunnel.

Even though we could ensure that users pick from the right box by using Augmented Reality visualizations, we could not yet benefit from Augmented Reality visualizations to ensure that the user also picks the right amount.

System

We found several usability problems in the Pick-by-Vision system. The main problem was that the button could accidentally be pressed twice and thus order lines were missed. Unfortunately we did not provide the functionality to go one step back. We additionally could observe that users performed faster with both types of Pick-by-Vison systems, than with Pick-by-Paper. However we could not show benefits for Pick-by-Vision (AR) over Pick-by-Vision (2D).

In this short test, users did not complain about being strained by the Pick-by-Vision system. Longer tests are required to get evidence for this.

Evaluating AR/ Lessons Learned

It is quite expensive to carry out an experiment in a real scenario. The test exposed the fact that the Meta Visualization is far from suitable for the real scenario despite testing well in our lab setup. Furthermore we could elicit quite important user requirements by letting real workers perform real tasks with our somewhat mature prototype. We furthermore found a tendency that users without previous experience with 3D user interfaces performed relatively poorly. In hindsight, we should have probably trained them more extensively in the try-and-ask phase and should look in future experiments for their long term performance.

5.7. A Stress Test and Concluding the Final Visualization

The previous experiment showed that Pick-by-Vision works in general and seems already to be comparable to the established Pick-by-Paper system. For this reason, we decided to test our Pick-by-Vision system in an in-depth evaluation over a "longer" period of time. We wanted to get insights into how learning effects affect errors and performance. Particularly when comparing our error rates with common error rates (see [140]), we could see a lot of potential for learning effects to influence the results after a longer period of use. Furthermore, we knew from Kampmeier et al. [76] that long-term use of an HMD can be a strain for the user resulting in headaches, eye fatigue and discomfort.

Objectives		Testing Pick-by-Vison (AR) over a longer period in a real setup and measuring strain in an objective way (heart rate variability analysis) .
Experimental Setup	Hypothesis	Users make less errors and are faster with Pick-by-Vision (AR) than with Pick-by-Paper. Pick-by-Vision produces more strain.
	Independent Variables	Type of Picking System: Pick-by-Vision (AR), Pick-by-Paper
	Dependent Variables	Time, Failure, (subjective and objective) Strain with different methods
	Test Method	Formal Experiment, within-subject design, warehouse 3: a real warehouse with 2 aisles and real items, manual user input: click-turn-wheel
Results	AR - Visualization	The Tunnel Visualisation was almost perfect (except FoV 4).
	System	Systems to prevent from jumping over articles, or picking the wrong amount. Some users have real problems using a HMD.
	Evaluating AR / Lessons Learned	Team of evaluators, intensive (unconventional) observing strategies can give a lot of insights - Muppet Show Balcony

Figure 5.32.: Overview of the Stress Test.

To this end, we designed a new experiment and thereby tried to reproduce the results of a strain test from Tümler et al. [146] (see also Sec. 2.2). As the execution of such a study goes along with significant expense, especially if users have to do real tasks, we had to set some limitations. In this respect, we decided to use only a few participants and follow a rather in-depth qualitative approach, rather than a quantitative one. Furthermore, we expected the Pick-by-Vision (AR) system to deliver more new findings than the Pick-by-Vision (2D) system. Finally, we set up an experiment comparing our Pick-by-Vision (AR) system in a "longer" study with a common Pick-by-Paper system. In Fig. 5.32 an overview of the expriment is given. This experiment was executed cooperatively [127]. In particular the analysis of the actual strain parameters, which is described in this chapter, was not performed by the author himself.

5.7.1. Objectives

The main intention of this experiment was to see what happens, when people use our Pick-by-Vision(AR) system over a longer period of time. We wanted to verify our visualization in general. Furthermore, we were interested in the influence of learning effects and fatigue on the logistic figures (time and error) and in the strain produced by the Pick-by-Vision(AR) system.

5.7.2. Experimental Setup and Hypotheses

For our testing environment, we used the same environment (warehouse and hardware) as in the previous experiment (Sec. 5.6). The setup and execution is described below.

Independent Variable: Type of Picking System at 2 Levels

The only independent variable of this experiment was the type of the picking system, which was used at the two levels: Pick-by-Vision (AR) or Pick-by-Paper.

In the previous experiment (Sec. 5.6) we established that the Tunnel visualization of the Pick-by-Vision (AR) system is not yet perfect, in particular when the user is turned further away than about 80 degrees. For that reason a new or adapted Meta Visualization for larger angles is required. We decided against using a new type of "Meta-Meta" Visualization such as additional Compass-like Arrows since this use of different presentation styles seemed to be more confusing than useful. Instead we improved the Tunnel animation, allowing the closest ring of the Tunnel to gradually slide away from the center of the display to move to the border. Using this animation scheme, users always see at least a small part of the Tunnel, helping them understand in which direction they have to turn in order to see more. The exact behaviour of the Tunnel in this Pick-by-Vision (AR) 2.0 system is described in Sec 6.1.

Test Method, Dependent Variables & Hypotheses

We designed a within subject experiment in which every test participant had to work with each technology for two hours and the start sequence of each technology was permuted. A single test session for one technology lasts about four hours. This includes the actual two hours picking, pre- and post-tests, recovery (resting) phases and up to half an hour try-and-ask introduction to the technology and a structured interview at the end of the test. Each test session for every test participant was executed on two different days at the same time of day, to get comparable results for the strain parameters. Within the two hours, test participants should fulfill at maximum 100 orders (with ∅2.9 order lines per order and ∅2.2 items per order line). Three boxes contained wrong articles. The wrong articles were supposed to be discovered by the users during the picking, by matching the article numbers printed on the item with the one presented by the picking system.

The analysis of user strain was done in the same way as it was in the experiment by Tümler [146]. This means that we analyzed the heart rate variability (HRV), and an EZ-Scale questionnaire and a discomfort questionnaire. In addition we again used the NASA TLX. The test participants wore a "Polar RS 800 CX Multi" pulse recorder for the analysis of the HRV. The software used to retrieve the data from the recorder was "Polar ProTrainer 5". Even if there was no difference in strain in the previous evaluation (Sec. 5.6), we expected the Pick-by-Vision system to produce a higher strain in this experiment. This was because in this test the participants had to work with the system for two hours.

To this end, we set up the following hypothesis which will be evaluated on the measurements obtained (HRV, EZ-Scale, NASA TLX and a discomfort questionnaire):

$H_{1-Strain-Exp7}$: Picking supported by Pick-by-Vision (AR) systems produces more strain than picking with Pick-by-Paper.

The Pick-by-Vision (AR) system already showed good results in the previous evaluation (Sec. 5.7) and moreover we expect learning effects from using the AR system over a longer period of time. Because of that, we set up the hypotheses that the Pick-by-Vision (AR) system should perform better according to the logistics figures:

$H_{1-Error-Exp7}$: The error rate is lower or the same with the Pick-by-Vision (AR) system compared to Pick-by-Paper.

$H_{1-Time-Exp7}$: The order picking time with the Pick-by-Vision (AR) system is better compared to Pick-by-Paper.

Observation Strategies To get deep insight into the experiment, we utilized several assistants besides the actual investigator, mainly to help with checking and sorting the picked articles. Moreover, we created the role of a special observer, who was placed on an observation deck a few meters above the actual experimentation area. This physical position had three benefits: the observer could perfectly observe, was not noticed by the test users (as if behind a traditional half-mirror) and, furthermore, was allowed to give comments to the test participants and even to interrupt the expriment. The observer watched the users' behaviour by mainly focussing on user strategies (how the user handles the system or how the user tries to conquer usability problems). To this end, the observer tried to analyze each ever-so-small step/movement a user did. While observing the user, the observer noted questions for an interview after the test. In cases where the user was observed to have severe trouble, in particular leading to unnecessarily bad performance, the observer could intervene in the experiment. We stopped the time during the breaks and solved the problem by giving a comment or e.g. adjusting the HMD. We think this policy falsifies the results less than not interrupting the user. After the test the users were confronted with the observations in a semi-structured interview. Such interviews typically lasted between fifteen and thirty minutes. The interviews were recorded using a tape recorder.

Later we named this way of observing experiments: *Muppet Show Balcony*.

5.7.3. Results

We had 8 users (4 male/4 female) between the age of 18 and 37 (mean age: 26, std dev: 5.37) performing the test. The test participants were, again, from all around the university, friends and as well as two professional order picking workers. Each of them got a wellness voucher as reimbursement.

Analysis of User Strain

The objective user strain was measured by analyzing the HRV data from the pulse recorder. In general, it can be said that the frequency of the heart rate is not an indicator for the stress level, which is better demonstrated through the variability in the heart rate. A high variability in heart rate indicates a low stress level, whereas a constant variability in the heart beat is an indicator for stress [146]. For the analysis of HRV the standard deviation (SD) was chosen as a parameter of the time domain which is used as a marker for short term changes of the sympathetic and parasympathetic nervous system indicating changes in user strain. Fig. 5.33 shows the development of SD through the test phases. The analysis of SD does not show a significant difference (Wilcoxon Test) between Pick-by-Paper and Pick-by-Vision (AR). This means that no higher physiological strain can be seen between both systems.

The EZ-Scale (Fig. 5.33) describes the subjective state of well-being of a person by Stanine values. Altogether changes in different factors can be seen for both systems indicating a rise in strain. Significant differences before and after the test can be found (Wilcoxon Test) for the Pick-by-Vision (AR) system in a reduction of activation ($p = .020$), rise in fatigue ($p = .027 < 0.05 = \alpha$) and decreasing relaxation ($p = .046 < 0.05 = \alpha$). For the paper list the willingness for exertion is reduced ($p = .039 < 0.05 = \alpha$) as well as social acceptance ($p = .034 < 0.05 = \alpha$) and sleepiness ($p = .041 < 0.05 = \alpha$). These data show that both systems result in a significant increase of strain but influence

Björn Schwerdtfeger

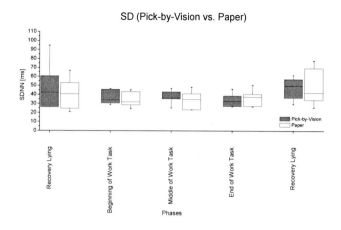

Figure 5.33.: Development of SD through the test phases.

different factors of personal well-being. This does not mean that one system produces more strain than the other.

The discomfort questionnaire asked for current physical complaints and mostly focused on the visual system of the user. The analysis of the data revealed significant differences between Pick-by-Vision (AR) and paper list only for the factor "headache" ($p = .034 < 0.05 = \alpha$, fig. 5.34).

The NASA TLX shows a task load of 87.81 for Pick-by-Vision (AR) and 71.98 for the paper list. However, there is no significant difference (ANOVA, $\alpha = 0.05$) (see Fig. 5.35). Compared to the previous experiment (Sec. 5.6), both values are higher, even though the actual task was the same. We lead this back to the fact that two hours of manual work without a break can be straining in general.

Overall the measured strain values for (HRV, EZ-Scale, discomfort questionnaire, NASA TLX,) show an increased strain for both systems Pick-by-Vision (AR) and Pick-by-Paper. The single relevant measurable difference is that the Pick-by-Vision (AR) produces headache. Only for the latter reason, we have to be careful with accepting the null hypothesis $H_{0-Strain-Exp7}$, that the use of Pick-by-Vision does not produce more strain than Pick-by-Paper.

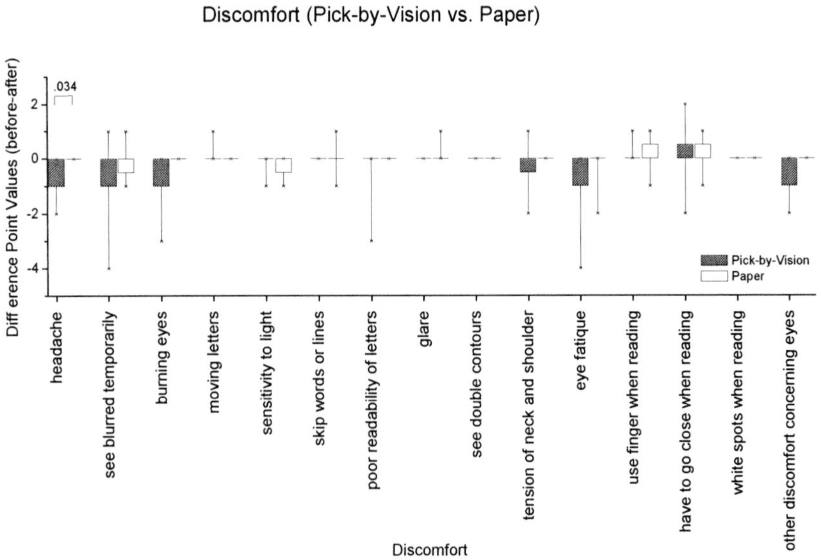

Discomfort (Pick-by-Vision vs. Paper)

Figure 5.34.: Result of discomfort questionnaire comparing differences between Pick-by-Vision (AR) and paper list before and after testing

Logistics Figures

The results of the logistics figures, error rate and time, will be discussed in the following paragraphs.

Error Rates   In this two hour test series the error rate for the Pick-by-Vision (AR) was higher than for the paper list. With Pick-by-Paper 5.873 order lines were accomplished and 29 errors were made, which results in an error rate of $(1.37\% \pm 1.09\%)$. With Pick-by-Vision (AR) 5.519 order lines were accomplished and 58 errors were made, which results to an error rate of $(3.03\% \pm 2.24\%)$. We corrected the value of the Pick-by-Vision system to $2.34\% \pm 2.08\%$, as users had some problems with the adjusting knob. The knob was not suited for long term use, and users sometimes pressed the button twice accidentally, and informed us of the mistakes rather than using the go-back function provided by the system.

From the mathematical point of view we should accept the null hypothesis $H_{0-Error-Exp7}$, that users made more errors with Pick-by-Vision (AR) than with Pick-by-Paper, but we

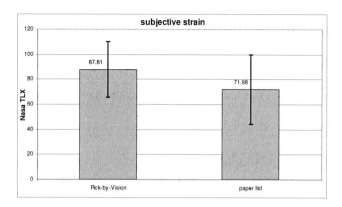

Figure 5.35.: NASA TLX for both technologies. Values can be between 0 (no task load) - 100 (full task load).

should also look at the reason behind it. The structure of the picking errors is shown in Fig. 5.36. With the paper list 14 errors were based on the wrong amount, seven on the wrong article and six on a missing order line. There were also two processing errors because the users forgot to sign the list.

With Pick-by-Vision (AR) the wrong amount was also the most frequent error (23). Despite discounting errors caused by the systematic system error, there are still 15 missing order lines. For example users clicked the button when they entered the aisle. This was not necessary and the first order line in this aisle was then missing. The process flow of the Pick-by-Vision (AR) system has to be improved in this case.

For AR the picking of wrong items is the most interesting error. The users picked five wrong articles, three of which were due to the wrong items we had put in. These three errors were not recognized by the test participants because they did not match the article number. The test participants did, however, pick from the right storage compartment. Therefore this error does not directly depend on the AR visualization.

One user picked two items of one order line correctly and one item from the storing compartment beneath. The user did not recognize the error because the items had the same shape. A more precise inspection of this leads to the following explanation: The user first took two items (three were not possible at the same time because of the weight). When the user took the third item, he did not move his head back to see the augmentation highlighting the box, but rather saw the box, while picking, only in the corner of his eye.

One user picked an item far away from the actual box. As we had to refill the warehouse manually and could not be 100% certain that all articles were placed correctly we think this mistake was due to an incorrect replenishment.

When we compared the distribution of errors over the time, we determined that the errors for the Pick-by-Vision (AR) system decreased during runtime and increased for the paper list.

Finally, we can state that the main reason for the bad performance of the Pick-by-Vision system is because it did not prevent users from picking the wrong amount or from accidentally skipping over an order line. Pick-by-Vision (AR) always helped to find the correct box.

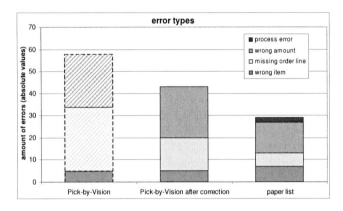

Figure 5.36.: Kinds of picking errors for both technologies before and after the correction of the systematic errors.

Time   Test participants were supposed to work with both techniques for two hours. However, some test participants fulfilled the 100 given orders in less time, some test participants had to take a break and we had a few technical breakdowns. Using the Pick-by-Vision (AR) system, users performed on average 123 (StdDev: 30) order lines per hour. When taking the interruptions into account, this amounts to 145 (StdDev: 39) order lines per hour. Using the Pick-by-Paper system, users performed on average 133 (StdDev: 27) order lines per hour, which does not change a lot, when taking the interruptions into account: 134 (StdDev: 27) order lines per hour. Even though the result (Fig. 5.37) shows a benefit of about 7.0% for Pick-by-Vision (AR), it is not significant (WIL, $p = .401 > 0.05 = \alpha$). Thus from the mathematical point of view we can not reject the null hypothesis to accept $H_{1-Time-Exp7}$.

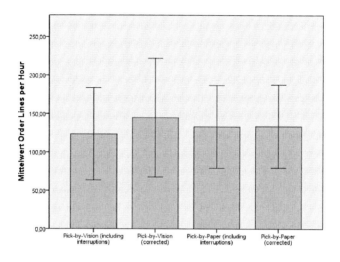

Figure 5.37.: Mean values of the order picking performance for both technologies with and without regarding the interruptions.

Problems using the HMD

There were several problems directly related to the HMD. Two users complained about headaches, which they traced back to the HMD headband. Three users complained about pressure in their eyes. Both observations were also made by [76, 41]. Two of our test participants had serious problems focussing on the HMD and stopped regularly during the experiment to try to focus the display. One of them even needed a 15-minute break because otherwise "the eye would have jumped out". We thought about psychological reasons, such as being afraid of the technology, but the test participants had a high affinity with new technologies. The other test participant said that as long as the background was in the focal distance she could focus on the HMD. We tested both test participants using a Landolt-C-Ring test and both had neither perfect nor particularly bad eyesight.

In general people said that they occasionally needed some time to focus on the HMD. However, test participants only complained about reading the text from the HMD, which was, in our opinion, displayed in an appropriate size (compare Fig. 5.27). None of the test participants had problems seeing the actual 3D augmentation. We reason that one does not need to focus on the 3D augmentation to be able to interpret and work with it. On the other hand, 50 % of the test participants did not have problems with the focus.

Other Observations

Against our expectations, the real picking workers performed slow compared to the other test participants. This has several reasons: they tried to do an error-free job because they know about the consequences of falsely delivered items. They placed all items in an organized way with the label side up in the sump, to have a fast overview before delivery, while other test participants just threw the items in the box. Furthermore, we realized that they behave like in their real job, where they actually do not work in a rush, as they usually carry out tasks for eight hours a day.

Besides this observation, we found out that there are two groups of HMD users. One group works in parallel: they make user input while walking. Members of the other group stop walking for each user input. Even when we interrupted people to suggest that they could work in parallel, they did not change their behaviour. For that reason, that group performed slower than those who worked using the parallel method.

5.7.4. Conclusions

Even if it does not look like it on the first view, the results of this first endurance test are quite satisfying: Test participants performed slightly faster using the Pick-by-Vison (AR) system. We could identify several issues of the Pick-by-Vision (AR) 2.0 system, which slowed down the usage of the system in this experiment: bad input device, too many system states to click through, no direct back button, late display of the aisle to go to, partial occlusion of the text by the augmentation and imperfect guiding by the Augmented Reality visualization. We could address most of these usability issues in the Pick-by-Vision (AR) 3.0 system, which we developed after this experiment.

System

According to the error rate, test participants performed three times worse than in the first evaluation. Even if the Augmented Reality visualization seems to be a perfect indicator for the right box, test participants could and did make other errors. Mainly they failed to check the article numbers or the amount. We traced this back to the fact that they had problems focusing on the text in the HMD, and some of them definitely ignored it. Furthermore, users probably grew tired and were not concentrating anymore. Those errors can be avoided by the use of other mechanisms, for example a speech input [115]. In such systems users have to speak the amount and an additional error checking number to confirm the order line.

Another important result of this study is that the results of the HRV, EZ-Scale and discomfort questionnaire show no general differences in strain changes between Pick-by-Paper and Pick-by-Vision (AR) which is very similar to the first study of user strain by Tümler [146]. From the EZ-Scale we can see that both systems influence different subjective parameters. The results of the discomfort questionnaire point out that working with an HMD can result in a headache, even though the reason for that remains unclear. A further observation is that two users had significant problems reading from the HMD and some other users mentioned a little pressure or some irritations in their eyes (however, this was gone after 5 minutes of not using the HMD).

A possible explanation for the headache is the general wearing comfort of the HMD (sometimes the headband was just too tight). However, there are other possible reasons for the problems with the HMD. The use of the HMD requires a frequent change of focus between the virtual image on the HMD and the real world, which in particular happens when long article numbers are matched. A further cause can be the two hour usage of a monocular display, which results in a binocular rivalry [81], as both eyes see different things. Beside that, there is no knowledge about the influence of longterm use (on e.g. eye fatigue and focus problems) of the direct retinal projection technology (of this HMD).

We can only draw the conclusion that further longterm assessments with different HMDs are needed. Those evaluations should be accompanied by an interdisciplinary team including psychological and medical experts.

Visualization

After a short introduction people were able to work well with the Pick-by-Vision system, except for one fact: the visualization in the FoV 4 area was too static. The Tunnel here did not provide any animated feedback to the test persons that would have helped them realize whether they were getting closer to the target or not: the Tunnel stuck motionlessly to the boarder of the display. In particular the following observation was helpful to find this usability problem: one user with a mental left/right weakness actually was confused by the static ring at the left/right boarder and moved into the opposite direction, due to translating the visual instruction into an abstract orientation concept and then switching directions when transforming it into physical motion (according to the user's own description). In reaction to that we had to find a dynamic Tunnel visualization for FoV 4, which will be presented and evaluated in the next and last experiment (Sec. 5.8).

Lessons Learned

Besides getting insights in the longterm usage of our system, we showed the useful application of an active observer, who had a complete overview and was allowed to intervene in the experiment. This prevented test participants from performing poorly and distorting the results, just because they wore the HMD in a bad way or moving the picking trolley inefficiently. We furthermore had several assistant evaluators, who allowed the chief evaluator to fully concentrate on observing the users. Furthermore those assistants discovered user behaviors the chief evaluator simply overlooked. We draw a conclusion regarding our philosophy of observing a few people in-depth rather than following an approach with a large number of users: if we had used more users in the tests, we probably would have had more statistically significant results, but the insights and conclusion would have largely been the same. However, we think thess in-depth evaluations with a smaller sample size need an experienced evaluator to produce valid results.

5.8. Verification of the Final Visualization

In the last experiment we found that the Tunnel visualization was not perfect in FoV4. It basically consisted of the first ring of the Tunnel, which is halfway outside the screen. The first rings stayed at this position during the whole area of FoV4 and thus did not provide feedback on the user's movement. To this end we had to find a way to animate the Tunnel such that it provides a positive/negative feedback on the direction in which the user moves. There are two different animation paradigms: which we call *contracting* and *growing*. By contracting, we mean the ring deforms horizontally into an elipse more and more the further the user turns away from the target box (see ring 2 in Fig. 5.38). Our intention behind this is to give it a rubber-like feeling. The alternative solution to the contracting is to make the ring grow in vertical direction (see ring 3 in Fig. 5.38). This was inspired by Baudisch et al. [8], who present off-screen locations of 2D maps by drawing a growing circle around the off-screen location. The size of this circle is adapted such that a fraction of it is always visible on the actual screen.

We discussed both alternatives with several experts and got preferences for the growing ring as well for the contracting one. This finally led to another user study. To this end we set up the warehouse from the first test-serieses (See Fig. 5.7), such that the user was placed in the middle and had a shelf to his left, right and front – the other warehouse was not available anymore. We observed nine users (from the hallway) in a within-subject design to assess both variants of the ring in FoV4. The within-subject design was chosen in this test also to compensate for preference effects – as users often prefer solutions they have tried first. The structure of the experiment followed our established approach: try-and-ask-introduction and 3 orders (with 9 articles) for each visualization. We hypothesized that it does not make a difference whether we grow or contract the ring, it is only important that the user gets a continuous feedback according to his motion.

5.8.1. Results

We could not measure any differences in task performance time, which is something we expected. However, we observed only nine users. What was more important was the

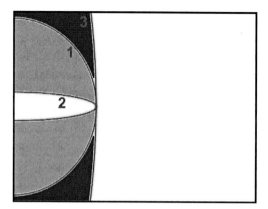

Figure 5.38.: This schematic view through the HMD shows the different possible behaviors for the ring in FoV4: 1) The ring in the start form, 2) the contracted ring 3) the growing ring. A comparison of the blue area with the green area shows that the optical difference between ring 1 and 3 is much smaller than between 1 and 2.

feedback of the users. We had two users who needed some time to get comfortable with AR and HMDs in general. They preferred the growing rings, as they liked the idea that there is something in the middle of the big ring. Two users had no preferences, they just looked on which side of the display the ring was located. The other five users preferred the contracting ring, because of two reasons. First of all, the contracting ring can give more detailed information about the magnitude of the rotation angle than the growing ring can. This is illustrated in Fig. 5.38. The green and blue areas show the possible optical differences between the full circle and the growing and respectively contracting ring. The green area for the contracting ring is much bigger than the blue area for the growing ring. This effect is amplified by the fact that the display of the HMD is wider than it is high. Furthermore they preferred the contracting over the growing, as they found the transit from FoV 3 to FoV 4 confusing with the latter type. They were in particular irritated that something grows when one moves further way. That is why they preferred the idea that something stretches rubber-like when one moves away. For that reason the final visualization (described in Sec. 6.1) makes use of the contracting ring and results in version number 3.0.

5.9. Final Presentation thus far

We presented the final Tunnel visualization to several users. For all earlier tests, we had to explain the visualization or at least give some small hints about how it works before users were able to use it. But with this new visualization, for the first time, we could just give users the HMD, start the demo, and people knew without further instructions what to do. One comment from an expert we got was: "How else should it work!"[1]

5.10. Conclusions

Doing research in industrial Augmented Reality is a challenging task, as industry partners want to know about the capability of a technology they invest in. Augmented Reality still has many unsolved problems but from time to time we need to compare our Augmented Reality applications with established technologies to assess its overall potential. We have shown how this is possible by carefully observing the reasons for bad performance of our prototypes.

There were still some errors made, using the Pick-by-Vision (AR) system. It was a perfect indicator, for the right box, but people picked the wrong amount or did not look at the article number to see that a wrong article was in the box. Those problems have to be conquered by other control mechanisms. Even though the Pick-by-Vision (AR) system was slightly better than the Pick-by-paper system, this benefit is typically not significant enough to introduce and apply the technology in industry. Thus further improvements and tests of our v3.0 system will follow.

Regarding user strain, we found that even though we have uncomfortable HMD headbands, a backpack to carry, and non-addressable display focal planes, our system did not cause a higher general user strain than the conventional paper list. Nevertheless, the discomfort questionnaire shows that improvements of the display devices are necessary to reduce the potential for headaches.

The problem remains that about 20% of the test participants had serious problems using the HMD. Brau and Fritzsche [76, 41] established the same problem for 20% of their users by also using the same Nomad HMD. In our case users did not have problems with the 3D augmentations, but just with reading the 2D text. Further investigations are necessary to find out if it takes some time for habituation or if it is directly related to the people or the type of HMD.

The two main results of this chapter are a mature visualization to support the order picking process with Augmented Reality and a long list of lessons learned for the evaluation of Augmented Reality user interfaces. The final Pick-by-Vision visualization makes use of a Box Finding and a Meta Navigation - a Frame with an adaptable Tunnel. This final visualization is fully presented in Chap. 6. The second conclusion of this series of

---

[1]Original statement in German: "Wie sollte es auch sonst funktionieren!"

evaluations – the lessons learned – resulted in Chap. 3, which presents a discussion about what to consider when evaluating Augmented Reality user interfaces. This chapter concludes with proposing a new usability engineering process for the development of usable and useful Augmented Reality systems.

The Final Pick-by-Vision Visualization

A description of the final Pick-by-Vision (AR) visualization. It consists of
a frame to highlight the box which is supported by an adaptable tunnel.

Make it as simple as possible. But not simpler.

ALBERT EINSTEIN

This chapter summarizes the details and provides the rationale for the final Pick-by-Vision (AR) visualization. The visualization is the conclusion of an intensive and iterative process, which was fully described in Chap. 5. It consists of a simple text-based coarse navigation for the Aisle Finding (Sec. 2.3.2) and a combined Tunnel and Frame navigation to support the fine navigation (Sec. 2.3.2) to find the actual box to pick from. Furthermore, we present some more details, which make the final Pick-by-Vision (AR) system convenient to use and successful.

6.1. The Final Visualization

This section describes our developed Pick-by-Vision (AR) visualization. It consists of the following three components: a) the Aisle Finding, b) the Box Finding, and c) the remaining information which is displayed as text.

6.1.1. Aisle Finding – Coarse Navigation

As we already described in the domain analysis (Sec. 2.3.2), the coarse navigation is not the critical part of the picking visualization. However, it has to be provided somehow. For that reason the system displays it as textual information, e.g: "Go to row 1.". Should it happen that the worker enters the wrong aisle, the fine navigation does not start, but the worker is instead notified and directed to the correct row. This makes the coarse navigation *zero error safe.*

There is one important aspect to consider. Probably the entire area of the warehouse is not equipped with a tracking system. Rather only the areas around the actual shelves provide tracking support, as was the case in our setup. However, rather than displaying no instruction, because of unavailability of tracking, our system assumes after some seconds of not receiving tracking data, that the worker is not in the tracking area and consequently not in the correct aisle. For that reason we then display the row to go to. This system behaviour is quite important, as we observed users (Sec. 5.6) going to the wrong row, and waiting for the tracking and instruction just to realize that they were in the wrong row.

6.1.2. Box Finding – Fine Navigation

The centerpiece of the Pick-by-Vision (AR) visualization is the Box Finding visualization, as presented in Fig. 6.1. It consists of a Frame visualization to highlight the box to pick from and the Tunnel as a meta visualization. In the first small artificial warehouse (see Fig. 5.7), we placed the Frame directly in front of the boxes. Later, in the industrial like warehouse (see Fig. 5.25), we did not highlight the box itself, but rather the labels directly under the box on the shelf. This was due to the simple reason that the boxes in the shelf are movable, whereas the labels stick to the shelf, and are thus always at the same position.

The Tunnel itself consists of thin rings which are aligned on a Bézier curve. In our case, this is a Hermetite curve [56]. It is described by a start point (a point some centimeters in front of the HMD), a start vector (the HMD's z-axis), an endpoint (the position of the box / or its label), and an end vector, which points perpendicularly out of the box. The individual rings of the Tunnel are rotated perpendicularly to their center point on the Bézier curve.

The rings are as thin as possible in order to obstruct the view of the real environment as little as possible. These thin rings make it possible that the Tunnel can stay in the field of view all the time and does not need to be faded out when the Frame is actually within the field of view. The Tunnel had to be faded out in earlier versions to make the Tunnel distinguishable from the Frame behind it. As a result the user can benefit from additional depth cues of the Tunnel (as described in Sec. 5.5.1), when the Frame is already within the field of view.

There is a problem when making the rings too thin: the rings which are further away can become blurred or even will not be shown because of the low resolution of the HMD. The solution to this problem is to dynamically increase the thickness of the rings, that are further away, such that they are always rendered clearly.

The different FoVs

As shown in the the first experiment in the real warehouse (Sec. 5.6), the Tunnel only provides a good guidance when the user is not turned away too far from the actual augmentation. The basic problem is that the Tunnel is not a perfect indicator for the direction to turn if the initial view is incorrect to a significant degree. In such cases, the user sees only one or two remaining rings of the Tunnel; the rest is outside the field of view. Obviously this is not enough information to indicate a clear direction and distance to turn.

When looking for a solution, we decided against using a new type of "Meta-Meta" visualization such as additional arrows, since this wealth of different presentation styles can confuse the user – we made an assessment by combining Tunnel and Compass-like Arrow. We experimented and discussed the idea of simply increasing the length of the start vector of the Bézier curve. An elongation of this start vector basically makes the Tunnel initially go straight away from the HMD, before starting to bend. Thus the user sees more rings of the Tunnel, even when being turned away further. This solution has a problem: it feels like having a long stick in front of the nose. Furthermore, this variant of the Tunnel quite often clashes with the oppositely facing shelf – which the user may be facing, before he turns around to the correct shelf. In addition to that, the bending of the Tunnel looks unnatural, when the start and end vector are too long.

The alternative solution, which we finally used, is allowing the closest ring of the Tunnel to gradually slide away from the center of the display, according to the angle between user and box and then to deform. To this end, we divided the behavior of the tunnel in four stages: Field-of-Views (FoV) 1-4 (compare Fig. 6.1). The angle alpha is calculated from the z-axsis of the HMD and the vector pointing from the box to the user (not to be confused with the angle which points straight out of the box). Alpha is the measurement for how far the user has to turn in order to have the actual augmentation in his viewing center. The different angles separating the different FoVs depend on the actual field of view of the HMD.

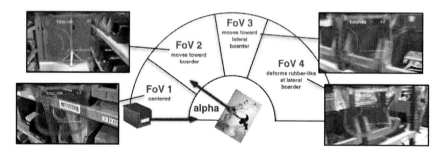

Figure 6.1.: The Figure shows the different behaviors of the Tunnel. It depends on the angle, alpha, between the vector pointing from the box to the user and the viewing direction of the user. Alpha thus indicates how far the user has to turn to have the current augmentation in the viewing center of the HMD.

FoV1: This FoV is about one third larger than the actual field of view of the HMD. In this FoV the Tunnel behaves like originally developed: it starts in the viewing center.

FoV2: Within this area, the start point of the Tunnel is geometrically interpolated between the viewing center and the maximal correction angle. This means that a sliding motion of the ring toward the boundary of the display occurs, in the direction of the target box. The maximum correction angle is such that the starting ring moves halfway out of the display. This means that at the maximum correction, the first ring can stick to any border of the display (top, bottom,left,right)

FoV3: Starting with FoV 3 we no longer adhere to a geometrically correct interpolation. At such angles, the user is already turned away quite far from the box, such that the geometrically correct and shortest path from the Tunnel goes mostly through the user's body or above his head. Thus in such situations, the user has to turn his body around first, before he can turn his head. This means the Tunnel has to guide the user to the left/ right, before the user can use the geometrically correct information within FoV2 and FoV1. To this end, within the FoV3, the end of the tunnel is interpolated between its geometrically correct position at the border to the center of the left/right border.

FoV4: In the first version of the FoV4 implementation, we simply stuck the first ring of the Tunnel halfway through to the left or right border, to indicate the direction to turn. Unfortunately, this did not provide continuous feedback to the users (compare Sec. 5.7.3) and did not show whether they were at the beginning (close to FoV3) of FoV4 or at its end (close to the other side). In particular the information as to how much the user has to turn to be in FoV3 was not given. To alleviate this problem, we animated the ring in this the fourth FoV, in a way that it shrinks vertically and seems to be stretched horizontally. In fact it transforms from a ring to an ellipse and thus feels like deforming in a rubber-like fashion (compare Sec. 5.8).

Transition left FoV4 - right FoV4: If the user turns further around into the wrong direction, the best path toward the box swaps from one side to the other; consequently a transition between the left and the right FoV4 occurs (or vice versa). In order to avoid erratic switching back and forth when the user remains close to this critical point, we buffered the transition. That means the user has to turn the head a few degrees more than 180 degrees to make the visualization change to the other side.

Other Possible Features

A design alternative to building the Tunnel out of individual rings would be to use something like a coil. According to [77], such Augmented Reality flight paths do not have to be connected, and thus no coil is needed. Yet, during all of our experiments some people, especially novices, would have benefitted from a coil like visualization in the early stage, since they did not understand immediately that the individual rings belonged together before having seen the whole tunnel. After having seen and briefly explored the whole Tunnel, they did not have a problem anymore. Thus a coil-like Tunnel could help to save some minutes of instruction, but not more.

A feature we implemented, but did not use, was to dynamically adjust the number of rings for the Tunnel, as the longer the Tunnel was the more frames were used. We tried to dynamically add/remove rings at both ends of the Tunnel. When we removed them at the end of the HMD, it looked like somehow walking through the tunnel; when we removed them at the end of the box, it looked like the Tunnel was disappearing in the shelf. Since both variants looked somehow unnatural, we switched this feature off. Furthermore, we realized that a static number of rings provides depth cues even when one does not see the whole Tunnel, due to the distance between the individual rings.

Furthermore , we tried animating the rings of the Tunnel dynamically like the lights of a runway at an airport. To this end, we increased/decreased the brightness of the rings sequentially. However, in some pre-tests, the users either did not realize this or they did not see a benefit in this runway light.

Version Numbers

The Pick-by-Vision (AR) Tunnel visualization with its behaviour was developed iteratively and thus was used within this work at different stages of development. We therefore assigned version numbers to the different stages of development. Version 1.0 consists basically of the implementation of FoV1. Version 2.0 makes use of all four FoVs, with the restriction that FoV4 was only visualized statically. In Version 3.0 all FoVs behave like they were described above.

6.1.3. Textual Information

Aside from the actual navigation, there is some other important information to be displayed, being the article number and the amount of items to pick. Both are displayed at the top of the display as text. During the endurance test (Sec. 5.7), we realized that in several situations, the textual information overlaps with the Tunnel or the frame, such that people had to move their heads to be able to read the text. To this end, we placed a black bar behind the text. However, this occludes the 3D information and keeps the text viewable at any time. This black bar appears to be transparent in an optical see-through display, as can be seen in Fig. 5.26.

The reason for not displaying the text of further items or even the whole list, is simply to keep the display clear. The decision to display only one item at a time is based on discussions and evaluations with the real workers. However, they requested to be able to display the whole list to get a fast overview on demand, upon special user input.

6.1.4. Accidental State Change

Another problem that we have to address is the robust state change, to prevent the user from accidently skipping an item. The visual switch to the next item can be overlooked, because the only things happening when the state changes are that the text (article number and amount) switches and that the Tunnel points to new direction - which happens instantly. To this end, we introduced an acoustic feedback to announce the state change via the speaker of the laptop. This feedback mechanism operated perfectly in all short tests. However, in the endurance test (Sec. 5.7), this was not enough as the users sometimes did nor register the (relatively loud) signal, because they either heard it to many times until they ignored it or they did not hear it because of the noisy industrial environment. Thus there is the need for a better visual feedback of the state change. This can be achieved by an animated visual transition of the items shown in the display, from one article number to the next. As the user does not continuously focus on the display, this visual transition possibly needs to be stretched over the whole screen, so the user cannot miss it. This is something which is required for the system. The perfect solution needs to be determined in further careful user tests.

Besides the extended visual feedback that we just described, and some external control mechanisms (like scanning the articles or the use of light barriers in front of the boxes), some other active support to the system is possible to prevent users from changing state accidently. We can make use of a lower time limit for giving a user input, preventing users from pressing the button twice, in quick succession. We have already made successful use of this technique, although only to detect such double clicks to correct the results in an experiment (compare Sec. 5.7.3).

6.2. Conclusions

In summary, there was a long way from our first setup in the lab over several evaluations and public exhibitions to the final visualization, which we can call an intuitive Augmented Reality visualization to support the order picking process.

We identified several facts to consider when designing such a visualization. First of all it must indicate, at any point in time and for any user position, how far users still have to move to have the actual augmentation in their field of view. If this indication does not exist, they move either too slowly or too fast. In the latter case, they often overshoot the mark and then have to turn back, in particular when the field of view is very small.

In addition, the visualization must always give a clear indication in which direction users have to turn. In doing so, attention has to be paid to the physical constraints of the user, i.e. the mathematically shortest way to turn is not always the best way to guide the user.

Furthermore, the Meta Visualization (in our case the Tunnel) should not be obtrusive. This is quite important as the user has to interact with his real environment. In this respect, the user should always be able to just ignore and see through the Meta Visualization in order to focus on other real or virtual objects.

There may be efficient alternative solutions to our Tunnel to guide a user wearing a tracked HMD in 4π steradians, but they probably do not yet exist. We have presented a mature solution, which is ready to be benchmarked with proposed alternatives in an endurance test.

Conclusion

An industrial Augmented Reality application was iteratively developed and evaluated. This included the design of a new metaphor for guiding users to objects outside their field of view. The lessons learned during all this evaluations led to a proposal of new ways to improve usability of AR systems.

Designing an object to be simple and clear takes at least twice as long as the usual way. It requires concentration at the outset on how a clear and simple system would work, followed by the steps required to make it come out that way - steps which are often much harder and more complex than the ordinary ones. It also requires relentless pursuit of that simplicity even when obstacles appear, which would seem to stand in the way of that simplicity.

T.H.NELSON, THE HOME OF COMPUTER REVOLUTION, 1977

For many years the development of the core technologies, particularly the tracking systems and display devices, was seen as the biggest problem of bringing Augmented Reality systems to real industrial settings. Even though there are a few workable core technologies that are currently available, developing actual Augmented Reality applications is still challenging. This work was started with the idea of simply using Augmented Reality visualizations to provide workers with a faster and better navigation when performing the order picking task. Although we were already experienced with Augmented Reality and had the benefit of expert advice, we developed this first system with the same naivity as probably most AR projects are started. We thought that as a) Augmented Reality is the most intuitive and thus best way of giving instructions for taks in a three dimensional world and b) when having tools and knowledge about the core technologies and calibration procedures available, it is an easy task to set up an Augmented Reality system to highlight objects in the real world. However, the development process of such a Pick-by-Vision (AR) system took several iterations of comprehensive user studies and redesigns, the invention and adaption of a visualization metaphor, the involvement of a huge amount of test participants and domain experts and several insights into how to perform usability studies. The result of proceeding by this long and iterative way are the four key contributions of this work:

- *Pick-by-Vision Prototype:* We developed a prototypical Pick-by-Vision system and thus did much more than demonstrating how Augmented Reality-supported order picking could look.

- *Tunnel Metaphor:* We developed and validated an intuitive Augmented Reality visualization metaphor to guide users with HMDs to objects outside their field of view.

- *Testing in the Real Environment:* We showed that the many problems can only be found and solved when testing the AR system repetitively under real conditions – which requires a relatively mature prototype.

- *New Usability & Utility Engineering Process:* We rethought the way in which we should evaluate and improve Augmented Reality systems.

Overall, we have shown that it takes a lot of resources and inventions to bring Augmented Reality to industrial setups. Several iterations of assessments and redesigns are required to make an AR system usable and useful and thus let users benefit from its "intuitive" and "natural" way of presenting information. That means a lot of problems in the user interface design have to be solved to design a competitive Augmented Reality system.

7.1. Contribution & Future Work

This section summarizes the achievements and unsolved problems of this work. Even though we have achieved a lot, there are still some problems to be conquered before Pick-by-Vision can be considered a state-of-the art technology to support the order picking process.

7.1.1. Pick-by-Vision

System

We have designed a convenient Augmented Reality based solution to support the order picking process. On the one hand, this required the solution of several usability issues. On the other hand we had to make the application useful, by finding and implementing the important user needs. We therefore started the development process with a domain analysis. However, we could only observe and elicit the important aspects when we already had a mature system and real order picking workers perform real tasks with the prototype in a realistic environment. As a result of this long and iterative process, we have developed Pick-by-Vision to a level where it can compete with established order picking solutions.

Now it is time to execute longterm assessments of the system in real environments. The only method to find out whether it is possible and useful to equip workers with tracked HMDs for a full work shift is to perform longterm assessments. Furthermore these assessments will give detailed information on the economic benefits of Pick-by-Vision. We do not yet know how a worker's picking performance changes over time, such as over the course of a full work day. We only observed benefits for Pick-by-Vision, when using it for two hours [114]. We also remain unaware of learning effects when using the system for several days.

Workers still did make some errors when using the Pick-by-Vision (AR) system. Although our visualizations indicated always the right box people picked the wrong amount, or did not match the article number to see that a wrong article was in the box. Those problems have to be conquered by other verification mechanisms.

Combined Tunnel & Frame Visualization

The main invention of this work was the Augmented Reality visualization to guide HMD-equipped workers in warehouses. It took several iterations of evaluations and optimizations to develop an intuitive, clear and usable Augmented Reality metaphor. The visualization consist of two elements: a clear indicator for highlighting the item location to pick from and a Meta Navigation, which guides the user to objects outside the field of view of the HMD. The picking visualization consists of a three dimensional Frame and the Meta Navigation of an highly adaptable ring based Tunnel. This Tunnel indicates at any time and for any position how far users still have to move to have the actual augmentation in their field of view. This takes the physical constraints of the user into account and is furthermore not obtrusive. The latter is quite important as the user has to interact with his real environment. Overall we invented a new Augmented Reality metaphor, which is ready to be benchmarked with alternative solutions in an endurance test.

7.1.2. HMD and User Strain

As found in our and other evaluations, some people get visual problems and headaches when using monocular HMDs based on retinal projection. These problems could not be traced back to a bad eyesight or to psychological reasons like being afraid of new technologies. We discussed this problem with several experts. Some argued that focusing on virtual images can be a problem in general or that the binocular rivalry (when using monocular displays) can be a problem. However, we could not find a definitive explanation so far. That means we need further long-term evaluations, with different

HMD-technologies, accompanied by elaborate medical and perhaps psychological observation, to get further insights.

Furthermore, we found that our system did not cause higher general user strain than the conventional paper list, even though we have uncomfortable HMD headbands, a backpack to carry, and non-addressable display focal planes. Nevertheless, the evaluation of the discomfort questionnaire shows that improvements of the display devices are necessary to reduce the risk of having headaches. The performance of long-term field studies is necessary to get more insights into the continuous load. We cannot test this in our setup as a real 8 hour workday with its interruptions (e.g. chats with colleagues, cigarette breaks, etc.) is different.

7.1.3. Usability & Utility Engineering

Usability and utility are probably the most important goals to be achieved when developing Augmented Reality based user interfaces. A high degree of usability and utility can only be ensured by an iterative assessment of the user interface in user based evaluations, as part of a well thought out design process. However, this requires Augmented Reality user interface experts, design guidelines, evaluation guidelines and experts in executing experiments – all of this is rarely available for Augmented Reality. Furthermore, Augmented Reality remains widely unknown, which means test participants have to get comfortable with the technology itself, before they can test the actual user interface. Additionally, the Augmented Reality design space is highly unexplored, which means that solutions and problems can be quite close to each other. All these aspects can have an unwanted influence on user behavior during the experiments, which is in most cases larger than the effect which results from the different controlled conditions. For that reasons, we suggest a rethinking of the way we currently evaluate Augmented Reality user interfaces. To this end we have to make use of new ways of observing users. Rather than treating every user in the same way, we have to consider the individual user, which requires us to actively intervene during the experiment and look for individual problems and perhaps guide the user or solve the problem before continuing. This active observation can be done in an ethnographic way: the evaluator and the user explore the user interface together or the evaluator sits somewhere distant from or above (on the "Muppet Show Balcony") the experiment and can intervene, if needed.

Probably the main mistake made is that experiments aim to show that an immature AR system performs better than an established technology by statistical hypothesis testing. In most cases, such a test fails, because we forget to shape the AR user interface. We should rather execute tests which will assist the redesign. Furthermore we should work with formal hypotheses, but see these hypotheses as exploratory research questions which define the goal of the experiment and the user interface design, rather than something which should be proved mathematically.

Finally, we proposed an adaption of the classical usability/utility engineering processes, by taking into account our lessons learned from all the experiments we executed within this work and in other projects. The process is a cost-effective but formal method which provides qualitative information for redesign. It aims to bridge the gap of the lack knowledge on AR design and evaluation guidelines by integrating a circular evaluation loop into classical processes, by which let designers and evaluators explore and conquer the design space iteratively and make them more and more experienced.

7.2. Final Words

In retrospect, the results of a properly executed empirical research experiment look obvious, more often than not. This is because we worked on something at the limit of what is known in our field and we thus had to invent something new. Finally we created a metaphor, which is intuitive, logical, ergonomic and thus obvious.

Augmented Reality has a huge but mostly unexplored design space and comes with a lot of other unsolved problems. Thus it takes a lot of effort to apply Augmented Reality to an industrial scenario and even more effort to make it perform better than an established technology, like paper based instructions. However, if we go this way of iteratively improving our applications, rather than trying to show (and fail) that Augmented Reality is better from the beginning, we have a good chance to reach the point from where we can make use of the real benefits of this technology.

While writing these final lines, there was already one logistics company trying to copy our Pick-by-Vision concept, another company had re-implemented an early version of our tunnel concept, and two companies (and one research institute) had discussed with us the possibility of setting up Pick-by-Vision in their facilities. It will be a promising future for industrial Augmented Reality...

List of Figures

2.1 A typical picking order ist 10
2.2 Common order picking technologies 11
2.3 Concept of a Meta Visualization 18
2.4 Related Meta Visualizations 18

3.1 New usability and utility engineering strategy 43

4.1 Fine adjustment of augmentation and shelves. 49
4.2 The Pick-by-Vision equipment. 51

5.1 Overview of all Evaluations. 58
5.2 Summary first experiment 59
5.3 Setup first experiment 61
5.4 Pathfinding Visualization 62
5.5 First Box Finding visualization 63
5.6 Summary second experiment 65
5.7 The experimental setup of the second generation. 66
5.8 Picking arrow and meta navigation in the second experiment. 67
5.9 Picking Frame and meta navigation in the second experiment. 67
5.10 First version of the Tunnel. 68
5.11 Screenshots from the test applications. 70
5.12 Error rates (according to the order) in the second experiment. 72
5.13 Times (according to the order) in the second experiment. 73
5.14 Average Times in the second experiment. 74
5.15 Correlation of possible predictors 1/2. 76
5.16 Correlation of possible predictors 2/2. 77
5.17 Correlation of the significant predictors. 78
5.18 Regression of the strong and independent predictors – coefficients. . . . 79
5.19 Two perspectives of the opaque Square-based Tunnel. 81
5.20 Summary third experiment 82
5.21 The semi-transparent Square-based Tunnel 84
5.22 The semi-transparent Ring-Tunnel. 84
5.23 Times (per order) in the third experiment. 87
5.24 Mean times in the third experiment 88
5.25 Setup of the "real" warehouse. 92
5.26 Pick-by-Vision view in the "real" warehouse. 94

5.27 The see-through Tunnel in the "real" warehouse.   95
5.28 Overview first experiment in the "real" warehouse.   96
5.29 Kinds of picking the overall errors.   99
5.30 Box plot of the times for experiment 6. 100
5.31 Average order picking time 101
5.32 Overview of the Stress Test. 103
5.33 Development of SD through the test phases. 107
5.34 Result of discomfort questionnaire. 108
5.35 NASA TLX for the stress test. 109
5.36 Kinds of picking errors in the stress test. 110
5.37 Average times in the stress test. 111
5.38 Schematic view of the Tunnel behaviour in FoV4. 115

6.1 The Final Pick-by-Vision Tunnel visualization. 121

Appendix

Configuration Files

A.1. Visualization

```
1  <?xml version="1.0" encoding="ISO-8859-1" standalone="yes"?>
2
3  <hmd>
4
5    <screen>
6      <device number="0" antialising="false"/>
7      <logFPS enable="false"/>
8      <position x="1024" y="0"/>
9      <size width="800" height="600"/>
10     <playSounds enable="true"/>
11   </screen>
12
13   <PickByVision2D>
14     <itemLocation="false">
15   </PickByVision2D>
16
17   <PickByVisionAR>
18     <!--Meta Visualizations-->
19     <guideArrow enable="false"/>
20     <rubberBand enable="false"/>
21
22     <wireframePath enable="true">
23
24       <visualsettings>
25         <curve yoffset="0.05"/>
26         <rotation enable="true"/>
27         <fovCheck enable="true">
28           <squeezeOnScreenBorder enable="true"/>
29         </fovCheck>
30         <animation enable="false" duration="1" goback="false"/>
31       </visualsettings>
32
```

```
33      <frame count="20" type="torus" faces="64">
34        <outerRadius>0.025</outerRadius>
35        <ringStrength>0.00035</ringStrength>
36      </frame>
37
38      <!--frame count="10" type="rectangle">
39        <width>0.035</width>
40        <height>0.035</height>
41        <depth>0.001</depth>
42        <strength>0.001</strength>
43      </frame-->
44
45    </wireframePath>
46
47    <!--Picking Visualizations-->
48    <itemArrow enable="false"/>
49    <itemFrame enable="true" width="0.125" height="0.040"
50          depth="0.005" lineStrength="0.005">
51      <offset><vector x="0" y="0" z="0.00"/></offset>
52    </itemFrame>
53    <itemBall enable="false"/>
54    <roadSign enable="false"/>
55
56    <!--Occlusion-->
57    <blockOcclusion enable="true" visible="false"/>
58
59    <!-- For Debugging -->
60    <centerCube enable="false"/>
61    <centerCube2D enable="false"/>
62    <centerAxis enable="false"/>
63    <ballGrid enable="false"/>
64    <nearestBoxArrow enable="false"/>
65
66    <!--Corners or edge around the screen-->
67    <corners enable="false"/>
68    <edge enable="false"/>
69   </PickByVisionAR>
70
71 </hmd>
```

A.2. Shelf Setup

```
 1 <?xml version="1.0" encoding="ISO-8859-1" standalone="yes"?>
 2
 3 <shelves>
 4
 5   <aisle name="Aisle_1/2">
 6
 7     <shelf>
 8
 9       <coordinates> <!-- origine of the shelf -->
10         <origin><point x="-1.675" y="1.8" z="-0.043"/></origin>
11         <toTheTop><vector x="0" y="0" z="1"/></toTheTop>
12         <intoBox><vector x="0" y="1" z="0"/></intoBox>
13       </coordinates>
14
15       <boxes> <!-- storage locations -->
16         <toFirstBox><vector x="0.115" y="-0.40" z="-0.03"/>
17           </toFirstBox>
18         <horizontalDistance number="7"><vector x="0.23" y="0" z="0"/>
19           </horizontalDistance>
20         <verticalDistance number="5"><vector x="0" y="0" z="-0.41"/>
21           </verticalDistance>
22       </boxes>
23
24       <size> <!-- occlusion object -->
25         <toStartPoint><vector x="-0.05" y="-0.39" z="-1.73"/></toStartPoint>
26         <dimension><vector x="3.37" y="0.4" z="2"/></dimension>
27       </size>
28
29     </shelf>
30
31     <shelf> <!-- other side of the shelf--> </shelf>
32
33   </aisle>
34
35   <aisle> <!-- another aisle--> </aisle>
36
37 </shelves>
```

A.3. Warehouse

```
1 <?xml version="1.0" encoding="ISO-8859-1" standalone="yes"?>
2
3 <items>
4
5   <item id="1">
6     <location>1A01</location>
7     <description>M6 x 40 Hexagon Bolt - SW10 DIN 933</description>
8     <image>hexagon_bolt_SW10_DIN_933</image>
9     <info>Box is heavy</info>
10    <partnumber>0057640</partnumber>
11  </item>
12
13  <item id="2"> <!-- next Item --> </item>
14
15
16 </items>
```

A.4. Orders

```
1 <?xml version="1.0" encoding="ISO-8859-1" standalone="yes"?>
2
3 <orders>
4
5   <order id="Order_1">
6     <item id="33" amount="3" />
7     <item id="133" amount="1" />
8     <item id="217" amount="1" />
9     <item id="152" amount="2" />
10  </order>
11
12  <order id="Order_2">
13    <item id="84" amount="1" />
14    <item id="93" amount="3" />
15    <item id="68" amount="2" />
16    <item id="299" amount="1" />
17    <item id="183" amount="2" />
18    <item id="244" amount="2" />
19  </order>
20
21  <!--and so on-->
22
23 </orders>
```

Bibliography

[1] Ajnaware Pty Ltd. *Sun Seeker Augmented Reality App for iPhone 3Gs.* `http://www.ozpda.com/`, Accessed December 2009. 4

[2] T. Alt. *Augmented Reality in der Produktion: Dissertation an der Otto-von-Guericke Universität Magdeburg.* Herbert Utz Verlag, 2003. 14

[3] M. Anastassova, J.-M. Burkhardt, and C. Mégard. User-Centred Design and Evaluation of Augmented Reality Systems for Industrial Applications: Some Deadlocks and Breakthroughs. In *Virtual Reality International Conference (VRIC)*, 2007. 24

[4] M. Anastassova, C. Mégard, and J.-M. Burkhardt. Prototype Evaluation and User-Needs Analysis in the Early Design of Emerging Technologies. In *Human-Computer Interaction. Interaction Design and Usability*, volume 4550 of *LNCS*. Springer Berlin / Heidelberg, 2007. 6, 24

[5] Apple Inc. *iPhone & Apples' App-Store (accessible via iTunes).* `http://www.apple.com/de/itunes/`, Last Checked: Dez 2009. 4

[6] R. Azuma. A Survey of Augmented Reality. *Presence: Teleoperators and Virtual Environments*, 1997. 3, 4, 34

[7] K. Backhaus, B. Erichson, W. Plinke, and R. Weiber. *Multivariate Analysemethoden - Eine Anwendungsorientierte Einführung.* Springer, 2000. 75

[8] P. Baudisch and R. Rosenholtz. Halo: A Technique for Visualizing Off-Screen Locations. In *Proceedings of CHI 2003, Fort Lauderdale, FL*, 2003. 114

[9] M. Bauer. *Tracking Errors in Augmented Reality.* PhD thesis, Technische Universität München, 2007. 3

[10] M. Bauer, B. Brügge, G. Klinker, A. MacWilliams, T. Reicher, S. Riß, C. Sandor, and M. Wagner. Design of a Component–Based Augmented Reality Framework. In *International Symposium on Augmented Reality ISAR*, 2001. 47

[11] A. Beu. *Style Guide für Augmented Reality Systeme.* `http://www.uid.com/arvika/`, Accessed December 2009. 52

[12] R. Bias and D. Mayhew. *Cost-Justifying Usability: An Update for the Internet Age, Second Edition.* Morgan Kaufmann, 2005. 21

[13] O. Bimber and R. Raskar. *Spatial Augmented Reality: Merging Real and Virtual Worlds*. A. K. Peters Ltd., 2005. 4

[14] F. Biocca, A. Tang, C. Owen, and F. Xiao. Attention Funnel: Omnidirectional 3D Cursor for Mobile Augmented Reality Platforms. In *CHI '06: Proceedings of the SIGCHI conference on Human Factors in computing systems*, pages 1115–1122, New York, NY, USA, 2006. ACM. 18, 19, 68, 83, 88

[15] J. Bortz. *Statistik für Human- und Sozialwissenschaftler*. Springer Medizin Verlag, Heidelberg, 6. edition, 2005. 36, 40, 62, 75

[16] D. Bowman, J. Gabbard, and D. Hix. A Survey of Usability Evaluation in Virtual Environments: Classification and Comparison of Methods. *Presence: Teleoperators and Virtual Environments*, 11(4), 2002. 34, 35, 36, 38

[17] D. A. Bowman, E. Kruijff, J. J. LaViola, and I. Poupyrev. *3D User Interfaces: Theory and Practice*. Addison-Wesley/Pearson Education, 2004. 25, 27, 52

[18] D. Bowmann and L. Hodges. Formalizing the Design, Evaluation, and Application of Interaction Techniques for Immersive Virtual Environments. *The Journal of Visual Languages and Computing*, 1999. 42, 44

[19] H. Brau, C. Ullmann, M. Duthweiler, and H. Schulze. Gestaltung von Augmented Reality Applikationen für Kommissionieraufgaben. In L. Urbas and C. Steffens, editors, *Zustandserkennung und Systemgestaltung Bd. 19*. VDI-Verlag, 2005. 14

[20] R. Brickenkamp. *Test d2 Aufmerksamkeits-Belastungs-Test, 8. Auflage*. Hogrefe, 1994. 39

[21] J. Brooke. SUS: a "quick and dirty" usability scale. In P. W. Jordan, B. Thomas, B. A. Weerdmeester, and A. L. McClelland, editors, *Usability Evaluation in Industry*, 1996. 39

[22] H. Brynzer and M. I. Johansonn. Design and performance of kitting and order picking systems. *International Journal of Production Economics*, 45:115–125, 1995. 10

[23] M. Burmester, M. Hassenzahl, and F. Koller. Engineering attraktiver Produkte - AttrakDiff[TM]. *In: J. Ziegler & W. Beinhauer (Hrsg.), Interaktion mit komplexen Informationsräumen*, 2007. 39

[24] A. Butz, M. Schneider, and M. Spassova. SearchLight - A Lightweight Search Function for Pervasive Environments. In *Pervasive 2004*, pages 351–356, 2004. 13

[25] J. Carroll and M. Rosson. Usability Specifications as a Tool in Iterative Development. *Advances in Human-Computer Interaction*, 1985. 35

Björn Schwerdtfeger

[26] C. Chen and M. C. (Eds.). Introduction to the Special Issue on Empirical Evaluation of Information Visualizations. *International Journal of Human-Computer Studies*, 2000. 28

[27] L. Damodaran. User Involvement in the system design process - A practical guide for users. *Behaviour & Information Technology*, 1996. 45

[28] W. Dangelmaier, B. Mück, M. Höwer, and W. Franke. AR Support for Picking. In *Simulation und Visualisierung 2005*. SCS European Publishing House, 2005. 16

[29] E. del Galdo, R. C. Williges, B. Williges, and D. Wixon. An Evaluation of Critical Incidents for Software Documentation Design. In *Proceedings of Human Factors and Ergonomics Society Annual Meeting*, 1986. 38

[30] D. Drasic and P. Milgram. Perceptual Issues in Augmented Reality. In *Proc. SPIE Vol. 2653*, 1996. 54

[31] J. Dumas, J. Redish. *A Practical Guide to Usability Testing, Revised Edition*. Intellect Books, 1999. 22

[32] A. Dünser, R. Grasset, and M. Billinghurst. A Survey of Evaluation Techniques Used in Augmented Reality Studies (TR-2008-02). Technical report, University of Canterbury, HITLabNZ, 2008. 6, 28

[33] A. Dünser, R. Grasset, H. Seichter, and M. Billinghurst. Applying HCI Principles to AR System Design. In *MRUI'07: 2nd International Workshop at the IEEE Virtual Reality 2007 Conference, Charlotte, North Carolina*, 2007. 23, 26, 27, 28

[34] F. Echtler, F. Sturm, K. Kindermann, G. Klinker, J. Stilla, J. Trilk, and H. Najafi. The Intelligent Welding Gun: Augmented Reality for Experimental Vehicle Construction. In S. Ong and A. Nee, editors, *Virtual and Augmented Reality Applications in Manufacturing*. Springer, 2003. 5

[35] K. Ericsson and H. Simon. Verbal Reports as Data. *Psychological Review*, 3, 87. 38

[36] L. Faulkner. Beyond the five-user assumption: Benefits of increased sample sizes in usability testing. In *Behavior Research Methods, Instruments & Computers*, 2003. 35

[37] S. Feiner, B. MacIntyre, and D. Seligmann. Knowledge-based augmented reality. In *Communications of the ACM, 36(7)*, 1993. 18

[38] J. Forlizzi and K. Battarbee. Understanding Experinece in Interactive Systems. In *Proccedings of the 2004 Conference on Designing Intractive Systems (DIS 04): Processes, Practices, Methods, and Techniques*, 2004. 25

[39] W. Friedrich. *ARVIKA-Augmented Reality in Entwicklung, Produktion und Service.* Publicis Corporate Publishing, 2004. 2, 5, 52

[40] T. Frimor. Augmented Reality Support in Logistics, Master's Thesis. Master's thesis, Technische Universität München, München, 2006. 4, 60

[41] L. Fritzsche. Eignung von Augmented Reality für den Vollschichteinsatz in der Automobilproduktion. Master's thesis, TU Dresden, 2006. 14, 111, 116

[42] L. Fu and G.Salvendy. Effectiveness of user-testing and heuristic evaluation as a function of performance classification. *Behaviour & Information Technology,* 21(2), 2002. 28

[43] J. Gabbard. *Researching Usability Design and Evaluation Guidelines for Augmented Reality (AR) Systems.* http://www.sv.vt.edu/classes/ESM4714/ Student_Proj/class00/gabbard/index.html, Accessed December 2009. 25, 28

[44] J. L. Gabbard, D. Hix, and J. E. Swan. User Centered Design and Evaluation of Virtual Environments. *IEEE Computer Graphics and Applications 19(6),* 1999. 27, 42, 43

[45] J. L. Gabbard and J. E. Swan. Usability Engineering for Augmented Reality : Employing User-based Studies to Inform Design. *IEEE Transactions on Visualisation and Computer Graphics,* 14, 2008. 28, 42, 43

[46] J. L. Gabbard, J. E. Swan, D. Hix, M. Lanzagorta, M. A. Livingston, D. Brown, and S. Julier. Usability Engineering: Domain Analysis Activities for Augmented Reality Systems. In *Proceedings of SPIE Steroscopic Displays an Virtual Reality Systems IX,* volume 4660, 2002. 23

[47] I. Gartner. *Gratner Hype Cycle 2008.* www.gartner.com, Accessed December 2009. 4

[48] Y. Genc, M. Tuceryan, and N. Navab. Practical Solutions for Calibration of Optical See-Through Devices. In *IEEE International Symposium on Mixed and Augmented Reality (ISMAR'02), October 2002,* 2002. 50

[49] Google Inc. *Android Market for the G-Phone.* http://www.android.com/ market/, Accessed December 2009. 4

[50] T. Gudehus and H. Kotzab. *Comprehensive Logistics.* Springer, Berlin, 2009. 9, 10, 13

[51] W. A. Günthner. *Neue Wege in der Automobillogistik: Die Vision der Supra-Adaptivität.* Springer, 1999. 2, 5, 13, 45

[52] W. A. Günthner, N. Blomeyer, R. Reif, and M. Schedlbauer. Pick-by-Vision: Augmented Reality unterstützte Kommissionierung. In *Abschlussbericht (AiF-FV-Nr. 14756 N/1)*, 2009. 2, 17

[53] S. Hart and L. Staveland. Development of NASA-TLX (Task Load Index): Results of Empirical and Theoretical Research. In P. A. Hancock and N. Meshkati, editors, *Human Mental Workload*, pages 139–183. North-Holland, Amsterdam, 1988. 39, 97

[54] M. Hassenzahl, M. Burmester, and F. Koller. Der User Experience (UX) auf der Spur: Zum Einsatz von www.attrakdiff.de. In *Usability Professionals 2008, German Chapter der Usability Professionals Association*, 2008. 25

[55] T. C. Haupt. *Persönlichkeitstyp und Stresserleben.* PhD thesis, Ludwigs-Maximilian-Universität München, 2006. 75

[56] D. Hearn and M. Baker. *Computer Graphics C Version.* Prentice Hall, 1996. 120

[57] F. Heider. *The Psychology of Interpersonal Relations.* John Wiley & Sons, 1958. 45

[58] S. Henderson and S. Feiner. Evaluating the Benefits of Augmented Reality for Task Localization in Maintenance of an Armored Personnel Carrier Turret. In *International Symposium on Mixed and Augmented Reality (ISMAR '09)*, 2009. 34

[59] A. Hickey and D. A. Requirements Elicitation and Elicitation Technique Selection: A Model for Two Knowledge-Intensive Software Development Processes. In *Proceedings of the Thirty-Sixth Annual Hawaii International Conf. on Systems Sciences (HICSS)*, 2003. 24

[60] A. Hickey and A. Davis. Elicitation Technique Selection: How Do Experts Do It? In *11th IEEE International Requirements Engineering Conference*, 2003. 24

[61] D. Hix and J. Gabbard. Usability Engineering of Virtual Environments. *Handbook of Virtual Environments: Design, Implementation and Applications*, 2002. 28

[62] D. Hix and H. Hartson. *Developing User Interfaces: Ensuring Usability through Product & Process.* John Wiley and Sons, 1993. 23, 24, 26, 28, 31, 32, 33, 34, 35, 36, 37, 38

[63] D. Hix, J. E. Swan, J. L. Gabbard, M. McGee, J. Durbin, and T. King. User-centered design and evaluation of a real-time battlefield visualization virtual environment. In *Proceedings of the IEEE VR'99 Conference*, 1999. 42

[64] D. Hix, J. E. Swan, T. H. Höllerer, Y. Baillot, J. L. Gabbard, M. A. Livingston, S. Julier, and D. Brown. A Cost-Effective Usability Evaluation Progression for Novel Interacrtive Systems. In *Proceedings of the Hawaii International Conference on System Sciences*, 2004. 28

[65] T. Höllerer, S. Feiner, and J. Pavlik. Situated Documentaries: Embedding Multimedia Presentations in the Real World. In *ISWC '99: Proceedings of the 3rd IEEE International Symposium on Wearable Computers*, page 79, Washington, DC, USA, 1999. IEEE Computer Society. 18

[66] K. Hornbaek. Current Practise in Measuring Usability: Challenges to Usability Studies and Research. *International Journal of Human Computer Studies*, 2005. 37

[67] K. Hornbaek and E. Froklaer. Two psychology-based usability inspection techniques studied in a diary experiment. In *3rd Nordic conference on human-compueter interaction (NordiCHI'04)*, New York, 2004. 33

[68] T. Howe and P. M. Sharkey. Identifying Likely Successful Users of Virtual Reality Systems. *Presence: Teleoper. Virtual Environ.*, 7(3):308–316, 1998. 68, 69

[69] M. Huber, D. Pustka, P. Keitler, E. Florian, and G. Klinker. A System Architecture for Ubiquitous Tracking Environments. In *Proceedings of the 6th International Symposium on Mixed and Augmented Reality (ISMAR)*, Nov. 2007. 4, 48

[70] A. Huckauf, M. H. Urbina, I. Böckelmann, L. Schega, R. Mecke, F. Doil, and J. Tümler. Perceptual issues in optical-see-through displays. In *submitted to eighth IEEE and ACM International Symposium on Mixed and Augmented reality*, 2009. 54

[71] A. Huckauf, M. H. Urbina, F. Doil, J. Tümler, and R. Mecke. Distribution of Visual Attention in Head-worn Displays. In *Proceedings of the ACM Symposium on Applied Perception in Graphics and Visualisation 2008 (APGV08)*, Los Angeles, California, USA, 2008. ACM. 54

[72] N. Jacobsen and B. John. Two cases studies in using cognitive walkthroughs for interface evaluation. Technical Report CMU-CS-00-132, Carnegie Mellon, 2000. 33

[73] R. Jeffries. User Interface Evaluation in the Real World: A Comparison of Four Techniques. In *Proceedings of ACM conference on Human Factors in Computing*, New York, 1991. 37

[74] C. Jones. *Patterns of Software Failure and Success*. Thomson, 1996. 21

Björn Schwerdtfeger

[75] A. H. Jorgensen. Thinking-aloud in user interface design: a method promoting cognitive ergonomics. *Ergonomics*, 33(4), 1990. 38

[76] J. Kampmeier, A. Cucera, L. Fritzsche, H. Brau, M. Duthweiler, and G. K. Lang. Eignung monokularer Augmented Reality – Technologien in der Automobilproduktion. In *Tagung der Deutschen Ophthalmologischen Gesellschaft "Augenheilkunde in der alternden Gesellschaft - Herausforderung und Chance"*, 2006. 14, 103, 111, 116

[77] L. J. Kramer, L. J. P. III, and J. J. A. I. andRandall E. Bailey. Pathway design effects on synthetic vision headup displays. In *Proceedings of SPIE – Volume 5424, Enhanced and Synthetic Vision 2004*, 2004. 18, 19, 68, 122

[78] S. Kujal. User Involvement: A review of the benefits and challenges. *Behaviour & Information Technology*, 22, 2003. 45

[79] M. Kurosu and K. Kashimura. Apparent usability vs. inherent usability: experimental analysis on the determinants of the apparent usability. In *Conference companion on Human factors in computing systems*, 1995. 25

[80] T. K. Landauer. *The Trouble with Computers: Usefulness, Usability, and Productivity.* MIT Press, 1995. 21

[81] R. Laramee and C. Ware. Rivalry and interference with a head-mounted display, 2002. 113

[82] D. Liepmann, A. Beauducel, B. Brocke, and R. Amthauer. *Intelligenz-Struktur-Test 2000 RIntelligence-Structure-Test 2000 R.* Hogrefe, Göttingen, 3rd, extended edition edition, 2007. 69

[83] R. Likert. A Technique for the Measurement of Attitudes. *Archives of Psychology (140)*, 1932. 39

[84] G. Lindgaard and J. Chattratichart. Usability testing: what have we overlooked? In *Proceedings of the SIGCHI conference on Human factors in computing systems*, 2007. 35

[85] S. Liu, D. Cheng, and H. Hua. An optical see-through head mounted display with addressable focal planes. In *7th IEEE/ACM International Symposium on Mixed and Augmented Reality (ISMAR)*, 2008. 54

[86] M. A. Livingston. Evaluating Human Factors in Augmented Reality Systems. *IEEE Comput. Graph. Appl.*, 25(6):6–9, 2005. 6

[87] Luca GmbH. *Pick-by-Point.* http://www.luca.eu/en/pick-by-point, Accessed December 2009. 11

[88] Metaio GmbH. *Junaio Augmented Reality Web 2.0 App for iPhone 3GS.* http: //www.junaio.com/, Accessed December 2009. 4

[89] P. Milgram and F. Kishino. A Taxonomy of Mixed Reality Virtual Displays. *IEICE Transactions on Information and Systems*, 1994. 4

[90] D. Mizell. Boeing's Wire Bundle Assembly Project. In *Fundamentals of Wearable Computers and Augmented Reality*, 2001. 5

[91] Mobilizy. *Wkitude Augmented Reality App for G-Phones or iPhone.* http://www. wikitude.org/, Accessed December 2009. 4

[92] R. Molich. Comparative Usability Evaluation. *Behaviour & Information Technology*, 23(1), 2004. 36

[93] B. Mück, M. Höwer, W. Franke, and W. Dangelmaier. Augmented Reality applications for Warehouse Logistics. In A. Abraham, D. Yasuhiko, T. Furuhashi, M. Köppen, O. Azuma, and Y. Ohsawa, editors, *Soft Computing as Transdisciplinary Science and Technology - Proceedings of the fourth IEEE International Workshop WSTST'05*, Advances in Soft Computing, pages 1053–1062. Springer-Verlag, 25 - 27 May 2005. 16

[94] I. B. Myers. *A guide to the development and use of the Myers-Briggs type indicator.* Press. Consulting Psychologists, Palo Alto, CA, 1985. 69

[95] J. Nielsen. Why you only need to test with 5 users. In *Jakob Nielsen's Alertbox*, *http://www.useit.com/alertbox/20000319.html*, Accessed December 2009. 35

[96] J. Nielsen and R. Molich. Heuristic evaluation of user interfaces. In *Proceedings of CHI 90*, 1990. 26, 27, 35

[97] J. Nielson. Finding Usability Problems Through heuristic Evaluation. In *Proceedings of ACM CHI'92 Conference on Human Factors in Computing Systems.*, New York, 1992. 37

[98] J. Nielson. *Usability Engineering.* Academic Press, Boston, MA, 1993. 21, 31

[99] J. Nielson and R. Mack. *Usability Inspection Methods.* John Wiley & Sons, 1994. 27

[100] D. Norman. *Emotional Design.* Basic Books, 2005. 25

[101] D. Norman and S. Draper. *User Centered System Design: New Perspectives on Human-Computer Interaction.* Hillsdale, NJ:Erlbaum, 1986. 23

Björn Schwerdtfeger

[102] D. A. Norman. *The Design of Everyday Things.* Basic Books, 1988. 26, 38

[103] D. A. Norman. *The Invisible Computer: Why Good Products Can Fail, the Personal Computer Is So Complex, and Information Appliances Are the Solution.* MIT Press, 2000. 21

[104] O. Oehme. *Ergonomische Untersuchung von kopfbasierten Displays für Anwendungen der erweiterten Realität in Produktion und Service.* PhD thesis, RWTH Aachen, 2004. 15, 53, 54

[105] D. Olsen. Evaluating user interface systems research. In *Proceedings of the 20th annual ACM symposium on User interface software and technology*, 2007. 28

[106] F. Ostendorf and A. Angleitner. *NEO-Persönlichkeitsinventar nach Costa und McCrae, Revidierte Fassung (NEO-PI-R).* Hogrefe, 2003. 69

[107] C. Owen, J. Zhou, A. Tang, and F. Xiao. Display-Relative Calibration for Optical See-Through Head-Mounted Displays. In *ISMAR '04: Proceedings of the 3rd IEEE/ACM International Symposium on Mixed and Augmented Reality*, 2004. 50

[108] K. Pentenrieder. *Augmented Reality based Factory Planning.* PhD thesis, Technische Universität München, 2009. 5

[109] A. Perer and B. Shneiderman. Integrating statistics and visualization: Case studies of gaining clarity during exploratory data analysis. In *Proc. of the SIGCHI conference on human factors in computing systems*, 2008. 32

[110] C. Plaisant. The challenge of information visualization evaluation. In *AVI '04: Proceedings of the working conference on Advanced Visual Interfaces*, pages 109–116, New York, NY, USA, 2004. ACM. 6, 28, 32, 40

[111] B. Rasch, M. Friese, W. Hofmann, and E. Naumann. *Quantitative Methoden 2. Einführung in die Statistik für Psychologen und Sozialwissenschaftler.* Springer, 2006 (2. Aufl.). 40

[112] R. Raskar, P. Beardsley, J. van Baar, Y. Wang, P. Dietz, J. Lee, D. Leigh, and T. Willwacher. RFID lamps: interacting with a self-describing world via photosensing wireless tags and projectors. *ACM Trans. Graph.*, 23(3):406–415, 2004. 13

[113] H. Regenbrecht, G. Baratoff, and W. Wilke. Augmented reality projects in the automotive and aerospace industries. *IEEE Computer Graphics and Applications*, 25(6), Nov.-Dec. 2005. 5, 27

[114] R. Reif. *Entwicklung und Evaluierung eines Augmented Reality gestützten Kommissioniersystems.* PhD thesis, Technische Universität München, 2009. 8, 9, 10, 11, 12, 15, 47, 53, 129

[115] R. Reif, W. Günthner, B. Schwerdtfeger, and G. Klinker. Pick-by-Vision comes on Age: Evaluation of an Augmented Reality supported Picking System in a real Storage Environment. In *6th International Conference on Computer Graphics, Virtual Reality, Visualisation and Interaction in Africa (Afrigraph 2009)*, 2009. 15, 52, 98, 112

[116] G. Reitmayr and D. Schmalstieg. Location based applications for mobile augmented reality. In *Proc. of the fourth Australasian user interface conference*, pages 65–73, 2003. 13

[117] A. A. Rizzo, G. J. Kim, S.-C. Yeh, M. Thiebaux, J. Hwand, and J. G. Buckwalter. Development of a Benchmarking Scenario for Testing 3D User Interface Devices and Interaction Methods. In *11th International Conference on Human Computer Interaction, Las Vegas*, 2005. 26, 28

[118] S. Robertson. Requirements Trawling: techniques for Discovering Requirements. *International Journal of Human Computer Studies*, 55, 2001. 24

[119] J. Rolland, A. D., and G. W. Towards Quantifying Depth and Size Perception in Virtual Environments. *Presence: Teleoperators and Virtual Environments, 4(1)*, 1995. 54

[120] J. P. Rolland, M. Krueger, and A. Goon. Multifocus Planes in Head-mounted Sisplays. *Applied Optics, 39(19)*, 2000. 54

[121] J. Rubin, D. Chisnell, and J. Spool. *Handbook of Usability Testing: How to Plan, Design, and Conduct Effective Tests, Second Edition.* John Wiley & Sons, 2008. 22

[122] B. Schwerdtfeger and M. Anastassova. How to Design and Evaluate Industrial Augmented Reality Applications. *Tutorial at International Symposium of Mixed and Augmented Reality (ISMAR)*, 2009. 24

[123] B. Schwerdtfeger, T. Frimor, D. Pustka, and G. Klinker. Mobile Information Presentation Schemes for Logistics Applications. In *Proc. 16th International Conference on Artificial Reality and Telexistence (ICAT 2006)*, November 2006. 4, 19, 47, 52, 60

[124] B. Schwerdtfeger and G. Klinker. An Evaluation of Augmented Reality Visualizations to Support the Order Picking. Technical report, Technische Universität München, Report TUM-I-08-19, München, 2008. 65, 97

[125] B. Schwerdtfeger and G. Klinker. Supporting Order Picking with Augmented Reality. In *Proc. of the seventh IEEE and ACM International Symposium on Mixed and Augmented reality*, September 2008. 65

[126] B. Schwerdtfeger, D. Pustka, A. Hofhauser, and G. Klinker. Using Laser Projectors for Augmented Reality. In *Proceedings of the 15th ACM Symposium on Virtual Reality Software and Technology (VRST)*, 2009. 11

[127] B. Schwerdtfeger, R. Reif, W. A. Günthner, G. Klinker, D. Hamacher, L. Schega, I. Bockelmann, F. Doil, and J. Tümler. Pick-by-vision: A first stress test. In *ISMAR '09: Proceedings of the 8th IEEE International Symposium on Mixed and Augmented Reality*, pages 115–124, Washington, DC, USA, 2009. IEEE Computer Society. 8, 11, 104

[128] M. Scriven. The Methodology of Evaluation. In R. Stake, editor, *Curriculum evaluation: American Educational Research Association monograph series on evaluation*, volume 1. Rand McNally, 1967. 31

[129] M. Scriven. *Evaluation thesaurus*. Newbury Park, Sage Publications, 4 edition, 1991. 31

[130] P. Sherman. *Usability Success Stories*. Gower Publishing, 2006. 22

[131] B. Shneiderman and C. Plaisant. Strategies for evaluating information visualization tools: multi-dimensional in-depth long-term case studies. In *Proceedings of the BELIV workshop, Advance Visual Interface Conference*, 2006. 21, 32

[132] B. Shneidermann and C. Plaisant. *Designing the User Interface: Strategies for Effective Human Computer Interaction*. Addisson Wesley, 5th edition edition, 2009. 21, 22, 24, 26, 27, 28, 31, 32, 33, 34, 40

[133] J. Sinn. *SPSS Guide - Correlation and Regression*. Wintrop University, http://faculty.winthrop.edu/sinnj/PYSCd 75

[134] J. G. Snider and C. E. Osgood. *Semantic Differential Technique: A Sourcebook*. Aldine Publications, 1969. 39

[135] C. Snyder. *Paper Prototyping*. Morgan Kaufmann, 2003. 24

[136] J. Spool and W. Schroeder. Testing Websites: Five users is nowhere near enough. In *In Proc. CHI 2001, Extended Abstracts*, 2001. 35

[137] D. Stone, C. Jarett, M. Woodroffe, and S. Minocha. *User Interface Design and Evaluation*. Morgan Kaufmann, 2005. 26, 27

[138] A. Sutcliffe. *User-centered requirements enginneering*. Springer-Verlag, 2002. 24

[139] J. E. Swan, S. R. Ellis, and B. D. Adelstein. *VR 2007 Tutorial: Conducting Human-Subject Experiments with Virtual and Augmented Reality.* 40

[140] M. ten Hompel and T. Schmidt. *Warehouse Management.* Springer, Berlin, 2004. 13, 103

[141] M. Tönnis and G. Klinker. Effective Control of a Car Driver's Attention for Visual and Acoustic Guidance towards the Direction of Imminent Dangers. In *Proceedings of the 5th International Symposium on Mixed and Augmented Reality (ISMAR)*, Oct. 2006. 35, 61

[142] M. Tönnis, C. Sandor, C. Lange, G. Klinker, and H. Bubb. Experimental evaluation of an augmented reality visualization for directing a car driver's attention. In *Proc. IEEE International Symposium on Mixed and Augmented Reality (ISMAR)*, 2005. 35

[143] N. Tracktinsky. Aesthetics and Apparent Usability: Empirically Assessing Cultural and Methodological Issues. In *Proceedings of the SIGCHI conference on Human factors in computing systems*, 1997. 25

[144] M. Tuceryan and N. Navab. Single Point Active Alignment Method (SPAAM) for Calibrating an Optical See-through Head Mounted Display. In *Proceedings IEEE International Symposium on Augmented Reality, ISAR'00, Munich, Germany*, 2000. 50

[145] J. Tümler. *Untersuchung zu Nutzerbezogenen und Technischen Aspekten beim Langzeiteinsatz Mobiler Augmented Reality Systeme in Industriellen Anwendungen.* PhD thesis, Otto-von-Guericke Universität Magdeburg, 2009. 8

[146] J. Tümler, R. Mecke, M. Schenk, A. Huckauf, F. Doil, G. Paul, E. Pfister, I. Böckelmann, and A. Roggentin. Mobile Augmented Reality in industrial applications: Approaches for solution of user-related issues. In *Proc. of the seventh IEEE and ACM International Symposium on Mixed and Augmented reality*, 2008. 16, 39, 103, 105, 106, 112

[147] C. Ullmann. Nutzerakzeptanz von Augmented Reality - Eine Fallstudie zum Vollschichteinsatz in der Automobilproduktion. Master's thesis, Universität Hamburg, 2006. 45

[148] J. Underkoffler and H. Ishii. Illuminating light: an optical design tool with a luminous-tangible interface. In *Proc. of the SIGCHI conference on Human factors in computing systems*, New York, NY, USA, 1998. 13

[149] VDI, Berlin. *VDI guideline 3590: Order picking systems*, 1994. 9

Björn Schwerdtfeger

[150] M. Walter, M.: Walter. MP3-Player verderben wie Frischobst. *Logistik*, 2/2008. 9

[151] C. Ware. *Information Visualization, Second Edition: Perception for Design*. Morgan Kaufmann, 2004. 53

[152] J. Whiteside, J. Bennett, and K. Holzblatt. Usability engineering Our experience and evolution. *Handbook of Human-Computer Interaction*, 1988. 35

[153] D. Wixon. Evaluating Usability Methods: why the current literature fails the practioner. *Interactions*, 10(4), 2003. 33

[154] M. Ziegler and M. Bühner. *Statistik für Psychologen und Sozialwissenschaftler*. Pearson Education, 2009. 40

[155] P. Zimbardo and R.J.Gerrig. *Psychologie*, volume 7. Springer, 2003. 29, 37, 39